Science
and
Sensibility

More praise for *Science and Sensibility*

"Professor Laidler has admirably accomplished several feats in this book. He explains the present state of our knowledge in the fields of hard sciences by presenting complicated phenomena in simplified and understandable terms, making that information useful to the nonscientist and the scientist alike. He shows how the humanity of the individuals who have led in our advancement of knowledge influenced the exploration and understanding of nature. But this book is far more that being a mere compilation of historical biographies. In elucidating the techniques of man investigating science, he has clearly shown that the coherence of the methods of science (from the investigation of the microscopic world to the galactic universe; from the inanimate to the living realm) involves the same type of thinking and reasoning used in other fields of endeavor.

"What a person discovers throughout this book in the investigation of nature is the immensity of creation, its complexity of organization based on statistical chance, and how slim is the chance that we are here at all. Laidler argues 'in favor of a rational philosophy of life based on observational and experimental evidence derived for the universe around us.'

"It is a most fascinating book!"

—Prof. John Meiser,
Department of Chemistry,
Ball State University

"Written for the nonscientist, Keith J. Laidler's *Science and Sensibility* will also have great appeal for scientists and their students. Using historical narratives and vignettes, Professor Laidler wisely and deftly explains what the physical, biological, and human sciences have accomplished and why scientists' methods of studying the natural world embody the kind of reasoning that should be used for making decisions in everyday life."

—Mary Jo Nye, Horning Professor of the Humanities,
Department of History,
Oregon State University

"Dr. Laidler has brought to *Science and Sensibility* a quality that is often lacking in technical works written for the general reader: he has managed to write an engaging survey of several fascinating fields of science, yet in a manner that conveys his respect for the intellect of the reader. Dr. Laidler utilizes the language of science to present the concepts of science with the precision and detail they deserve, but well within the perspective of the general reader, so that the ideas are communicated with straightforward elegance. Dr. Laidler connects the ideas of science to the innate logic of the intelligent reader."

—Cathy Cobb, author of *Magick, Mayhem, and Mavericks:
The Spirited History of Physical Chemistry*

Science and Sensibility

The Elegant Logic of the Universe

Keith J. Laidler

 Prometheus Books

59 John Glenn Drive
Amherst, New York 14228-2197

Published 2004 by Prometheus Books

Inquiries should be addressed to
Prometheus Books
59 John Glenn Drive
Amherst, New York 14228-2197
VOICE: 716–691–0133, ext. 207
FAX: 716–564–2711
WWW.PROMETHEUSBOOKS.COM

08 07 06 05 04   5 4 3 2 1

Library of Congress Cataloging-in-Publication Data

Laidler, Keith James, 1916–2003.
    Science and sensibility : the elegant logic of the universe / Keith J. Laidler.
        p. cm.
    Includes bibliographical references and index.
    ISBN 1-59102-138-3 (alk. paper)
    1. Science—Popular works. 2. Science—Methodology. I. Title.

Q162.L315 2003
500—dc22

2003018605

Printed in the United States of America on acid-free paper

To see a World in a Grain of Sand
  And a Heaven in a Wild Flower
Hold Infinity in the palm of your hand
  And Eternity in an hour.

William Blake, *Auguries of Innocence*, c. 1803

Contents

Preface

How should we cope with the vast amount of information, about science and other matters, that confronts us daily? People seem eager to acquire information, but are less concerned about what they should do with it. Many spend time exploring the information highway. This can be a useful and timesaving activity, but far from being an end in itself, information is only the lowest rung of the intellectual ladder. Above it there is knowledge, which results after our brains have carefully selected information in an appropriate way, and have processed it into a coherent point of view. Beyond that, we hope, is wisdom.

One of the themes running through this book is how we process information to arrive at knowledge of the complex world we inhabit. I begin with science, which today has such a powerful influence on our material lives. More important, it is transforming our culture, as we discuss in a little detail in chapter 6. For these reasons I think it useful for all of us to have a clear understanding of the methods and influence of science. That is not the same thing as saying that everyone needs an extensive scientific education. After all, we can enjoy and appreciate music without being able to play or even read a note of it. So it is with science: since science and technology so dominate our everyday lives, we should try to understand and appreciate what science is about.

To make this book accessible to nonscientists, I have given a brief and simplified view of science, emphasizing the relationship between its different branches and showing how they lead to a unified conception of our place in the universe. Chapters 2 through 5 are concerned with the basic ideas and conclusions of chemistry, physics, astronomy, geology, and biology. We will see how evidence from these fields leads to an elegantly logical and self-consistent picture of the formation and development of the universe and of life within it.

Chapters 6 and 7 deal with human culture, beginning with some of the pitfalls that we encounter in forming our opinions. Although I believe that we have free will, we are all to a great extent conditioned by our heredity and by the environment in which we were raised. In these two chapters I discuss how some of the important conclusions reached by scientific investigations can be useful in dealing with seemingly unrelated problems. One such conclusion is that because nature and nurture are not additive factors, it is meaningless to try to estimate their relative importance. Another is the importance of chance in everyday events. Still another relates to the significance of feedback, that is, when a process is occurring continuously, some of the consequences may, apparently paradoxically, influence earlier links in the chain of events.

The last chapter, "What Is Truth?" is mainly concerned with emphasizing that however sincerely we use logic and reason, some doubt must always remain.

Several different techniques have been used by scientific writers, a few of whom seem to think that a typical nonscientific reader is a person who cannot grasp the idea of a relationship between two quantities, who would be terrified by a mathematical equation, and who would think a diagram an affront to dignity. My ideal reader is rather different, since, for reasons I explain in the book, I think that science involves essentially the same type of thinking that is employed in the professions or in business. Of course, there are many details of science that are hard to follow unless one studies them deeply, but the general ideas of science are less formidable.

This book is particularly directed toward readers who have some competence in other fields but who happen not to have studied much science. I have not hesitated to make some simplifications, which will be obvious to my fellow scientists and which I hope they will forgive in view of the main objectives of this book. I have limited my discussions to those aspects of science that I think the nonscientist reader will find most interesting and helpful. In spite of the liberties I have taken, I believe that the somewhat impressionistic picture I have painted gives a reasonably correct idea of how and what we learn from science.

I have not hesitated to include a few chemical and mathematical equations, but all of them are simple, with full explanations. I cannot believe that readers are incapable of grasping what $E = mc^2$ means, or understanding an inverse-square-law relationship. I have included diagrams wherever I think they would be helpful.

The quotation at the beginning of this book may appear surprising, since William Blake rather disliked and despised science. He intended his lines to lead to mysticism, but today they could lead us toward a life of intellectual adventure, in which we do our best to understand the mysteries of the wonderful universe around us. The mystic accepts and enjoys these mysteries. The scientist enjoys them just as much, and perhaps more and more as the great veil of ignorance is gradually lifted by the advance of science. This book tries to sustain our sense of wonder as we explore our understanding of nature.

Acknowledgments

I am greatly indebted to many people who have helped me while I have been writing this book. Dr. Alan Batten, of the Herzberg Institute of Astrophysics, Dominion Astrophysical Observatory, gave me valuable advice on cosmological matters. Correspondence over the years with Dr. Walter Gratzer, of the Randall Institute of King's College, London, has helped me with my understanding of molecular biology. My many discussions with Professors John Holmes and Brian Conway of the University of Ottawa have always been helpful. Dr. June Lindsey and Dr. George Lindsey, both physicists, have critically read the entire manuscript and have made many constructive suggestions as to scientific content and style of presentation.

Since this book is primarily intended for readers who are not scientists, I needed help from people who would be willing to judge the book from that point of view, and give me appropriate advice. I am particularly indebted to Dr. Christine Davenport, who has made a deep study of history and law but whose education included little science, for reading much of what I wrote in early drafts and advising me as to its suitability for the general reader. Much thanks is also due to my editor, Linda Regan, for her careful scrutiny of the book and her always constructive criticisms.

My son Jim Laidler has prepared the line drawings in this book, and has been of great help with finding suitable portraits. Only one of the portraits included is copyright-protected, that of Edwin Hubble, and I am indebted to the Archival Services of the California Institute of Technology for supplying a formal portrait of him and granting me permission to use

13

it. For copies of the portraits of Francis Bacon, Albert Einstein, Max Planck, Neils Bohr, James Hutton and Joseph Black, and Gregor Mendel, I am grateful to the Annenberg Rare Book and Manuscript Library, University of Pennsylvania, and in particular, I thank John Pollack for his personal help. For the interesting portrait of Ernest Rutherford in his early years, I am specially indebted to Professor John Campbell of the University of Christchurch, and to members of the Rutherford family.

A Few Points
about Mathematics

In this book I have kept mathematics to a minimum and have not included many mathematical equations. Some mathematical equations, however, like Einstein's $E = mc^2$, can always help, provided that there is an adequate explanation. I have therefore not hesitated to use these and a few others.

Scientific Notation

From time to time we need to use very large or very small numbers, and it is useful to use what is called scientific notation to avoid writing out strings of zeros. Thus, instead of writing 1,000,000, which is one million, we write 10^6, which is ten multiplied by itself six times, or one followed by six zeros; we pronounce it "ten to the sixth power," or "ten exponent six," or just "ten to the sixth." Similarly, one billion, or 1,000,000,000, we write as 10^9. If we want to write 602,200,000,000,000,000,000,000 (which happens to be the number of molecules in eighteen grams, or a *mole*, of water), we write 6.022×10^{23}, which is a lot more convenient.

A similar notation applies to very small numbers. Suppose we are dealing with 1/1,000,000 (one millionth), which can also be written as 0.000001. To write it in scientific notation, we note that one millionth is $1/10^6$, and write it as 10^{-6}. In other words, it is one divided by six tens multiplied together (i.e., 1,000,000).

15

A simple way of determining the exponent in such a case is to see how far we have to move the decimal point to the right in order to get the number one (unity); for 0.000001 we have to move it six places to the right. The value of an important constant, called the Boltzmann constant, is 1.381×10^{-23}, and this means that it is 1.381 divided by one followed by twenty-three zeros.

An important point about scientific notation is that when the exponent changes by one, the number itself changes by a factor of ten. Thus 10^4 is ten times as large as 10^3, and 10^9 (one billion) is one thousand (10^3) times as big as 10^6 (one million). The number 10^{-23} is one thousand times smaller than 10^{-20}.

Exponential Quantities

Exponential quantities are used not just to represent numbers, but in another way. In chapter 6 we shall meet the quantity exp $(-E/k_BT)$, which is just another way of writing

$$e^{-E/k_BT}$$

We pronounce this as "e to the minus E over k sub-B T," or "exponential minus E over k sub-B T," By this expression we mean that the number e ($2.71828\ldots$,) is raised to the power of $-E/k_BT$.

An important thing to remember is that anything raised to the power of zero is one.

Logarithms

When we use a logarithm, we are doing the inverse of what we do in forming an exponential. We have seen that for 1,000,000 we can write 10^6. We can also say that the logarithm of 1,000,000 is six. This kind of logarithm of a number is just the power to which ten is raised to get that number.

This particular logarithm, equal to the power to which ten is raised, is said to be to the base ten, and is also called a *common logarithm*. Logarithm tables, formerly used to make calculations and now superseded by computers, employed common logarithms. To avoid ambiguity when we use common logarithms, we use the notation $\log_{10}$; thus $\log_{10} 1,000,000 = 6$.

In scientific work we more often use what we call natural logarithms, written as $\log_e$ or more often as ln. These are to the base e, which is the odd little number we used earlier, having the value $2.71828\ldots$. Natural logarithms are bigger than common logarithms by the factor 2.303. Thus, $\log_{10}$ 1,000,000 = 6 and ln 1,000,000 = 2.303 × 6 = 13.818.

A change in a logarithm means a larger change in the quantity measured. The pH scale is a common logarithmic scale, and an inverse measure of acidity. Thus, a solution of pH 2 is ten times as acidic as one of pH 3 (ten times because it is a common logarithm, to the base ten). The Richter scale, named after the American seismologist, Charles Francis Richter (1900–1985), is also a common logarithmic scale. It is based on the amplitudes of vibrations that occur in an earthquake, as measured by a seismograph. An earthquake of seven on the Richter scale, for example, has average amplitudes that are ten times as large as for a six on the Richter scale.

The Metric System

Many of the important relationships in science can be expressed and understood much more conveniently using the metric system. It has therefore become part of the language of science, and been adopted for general use in most countries. Even in countries that have not adopted the metric system, primarily the United States and Burma, the metric system is necessarily used in teaching science.

The metric system is primarily used here, usually with the Imperial unit given in addition. The following tables of conversions may be useful to those not familiar with metric units.

Length

	inches	feet	yards	miles
1 meter (m)	39.37	3.28	1.094	
1 centimeter (cm)	0.3937	0.033		
1 millimeter (mm)	0.0394			
1 kilometer (km)				0.6214
1.609 km				1

Speed

	km/hour	feet/s	miles/hour
1 meter/second	3.6	3.28	2.237
1 kilometer/s	3,600	3,280	2,237
1 kilometer/hr	1.000	0.911	0.6214

Mass

	ounces	pounds	Long ton = 2,240 lb. = 1,016.05 kg	Short ton = 2,000 lb. = 907.19 kg
1 gram	0.0353	0.002205		
1 kilogram	35.27	2.205		
1 tonne (metric ton, 1,000 kg)		2,204.6	0.9842	1.102

Temperature

The fundamental scale of temperature is the absolute or Kelvin scale, which is based on absolute zero. Some equivalents in the Celsius scale (very close to the Centigrade scale) and the Fahrenheit scale are as follows.

	Degrees Celsius	Degrees Fahrenheit
0 K	−273.15 (by definition)	−459
3 K	−270.15	−454
273.15 K	0	32
293.15 K	20	68
373.15 K	100	212
5,000 K	4,727	8,577
6,000 K (Sun's surface temperature)	5,727	10,341
10,000 K	9,727	17,541

Chapter 1
To Tell the Truth

Where is the wisdom we have lost in knowledge,
Where is the knowledge we have lost in information?
—T. S. Eliot, *The Rock*, 1934

If we are to live useful and fruitful lives, we must constantly make intelligent and prudent decisions. We need to decide which foods to eat, how much and what we should drink, where our children should be educated, and what political party (if any) we should support, to name but a few practical topics. In making these decisions we are faced with an overwhelming amount of information, coming from a variety of sources, and of varying reliability. Aside from weeding out all the misinformation, we must process the valid information—select from the enormous mass of information, and use our reasoning powers to reach a coherent and rational conclusion.

Besides having to make decisions on practical matters, we need to come to terms with the world around us. We live in times of constant change, which many people find bewildering. Standards of behavior are different from what they were a few decades ago. Some of us are concerned with what we perceive to be a lowering of standards. Religious leaders are disturbed by the decline in attendance at religious observances. Scientific discoveries leading to technological change, especially the great increase in the volume and speed of communication, are playing an increasing role in bringing about these cultural transformations.

To reconcile ourselves to the changing times, we must handle all of the information and outside influences that come our way and arrive at some rational points of view. We should all understand just what science is concerned with, and what it teaches us. In this book, I first give a very general idea of science, and then discuss our culture, including religious belief, in the light of what we have learned from science. We should first consider a few general points about the intellectual methods that are found to be most effective, not only in science but in everyday life.

We reach most of our opinions and beliefs by analyzing the information we have acquired in our lives. I say "most" rather than "all," because some of our conclusions, although they may be perfectly reasonable, do not result from any rational thought processes. Instead, they arise from what we call instinct or intuition, and some of our attitudes are implanted in our minds at an early age, in which case we find it hard to discard them. For example, those of us who prefer Bach to pop music do not do so as a result of much intellectual effort, but largely as a consequence of our backgrounds. This book is concerned with discussing how we can best use the available information so as to come to rational opinions and beliefs.

What do we mean by information? Information includes: perceived facts, misinformation, and meaningless statements. What is a fact? Strictly speaking, a fact is something that is true, but it is important to realize that the word *fact* nearly always means a perceived fact. For example, it would be reasonable for me to say that it is a fact that at this moment I am writing this book. If (as I hope never happens) I have to appear in court to testify that I wrote this book, my testimony would appear as perceived facts. If another person testified that he or she had written this book, that would be another set of perceived facts. We could not both be telling the truth. Perhaps this seems obvious, but in a recent Belgian court case an expert witness testified that two witnesses who told completely inconsistent stories as to the facts could both be speaking the truth! This kind of nonsense comes from a fashionable modern (or postmodern) sociological fad called "postmodern relativism," which we will ignore for the most part in this book except for a few exquisitely absurd examples of it in chapter 6.

Misinformation, since misinformation is something of which we are informed, even though the truth may have been distorted, sometimes deliberately. Misinformation is alarmingly prevalent; a search of the Internet quickly shows this. Also, much of what passes for information is meaningless—one only has to listen to politicians to know that there is a lot of that around. I regret that even some scholars and scientists make meaningless statements; the anthropologist Teilhard de Chardin was a past master of the art, and here is an example from his *The Phenomenon of Man*, published in 1955: "Everything does not happen continuously at any one moment in the universe. Neither does everything happen everywhere in it."[1]

The physicist Wolfgang Pauli had a nice way of expressing his contempt when anyone said something meaningless; he would say, "That's not even wrong." One is reminded of the timeless remark by film director Sam Goldwyn, "I won't recognize him—I won't even ignore him." Being "not even wrong" is what makes meaningless statements so insidious. If something is merely wrong, we can deal with it and try to put it right; if it is not even wrong, and the person expressing the opinion cannot be convinced that it is nonsense, how can we combat it? The great physicist Lord Rutherford used to emphasize to his students that, in scientific work, one must always have *some* theory. It does not matter if the theory is wrong as long as we do not hold on to it unreasonably. If our theory is wrong, the new results may put it right and nothing has been lost. But if we have no theory, or a meaningless theory, we are working in the dark.

This book is mainly concerned with perceived facts that are correct as far as we are able to discover. For practical purposes, we will presume a fact to be innocent until found guilty. An error, on the other hand, is a piece of information that can be proven wrong; however, sometimes a guilty fact is later proved innocent. A useful word in science is *data*, facts that have been obtained by observation or experiment and are therefore "given." *Datum* (plural: *data*) is a Latin noun meaning a "thing given."

There are serious problems in dealing with information. As a result of the enormous numbers of books and periodicals, and now the Internet, we always have vastly more information than we can manage. It has been estimated that the amount of information on the World Wide Web doubles every three months. If that continues, in ten years the amount will be a thousand billion times what it is now, and in twenty years it will be a million billion billion times. Before long there will be more bits of information than there are atoms in our computers. I may be wrong, but I doubt it will be practicable for computers to store that amount of information, so presumably something will be done to control the volume. In any case, it looks as if there will never be a shortage of information. The problem is knowing how to handle it.

First, we must determine which information is believable. How we do this depends on the type of information. Scientific facts, for example, may be checked by repeating experiments many times. Historical facts may be checked by going back to several sources. In this book we shall see a few examples of how we arrive at reliable information.

There is, of course, much more to dealing properly with information than merely distinguishing the reliable from the unreliable from the meaningless. Most important is to select relevant information intelligently. An important comment about selection was made over a century and a half ago by the great historian Lord Macaulay: "He who is deficient in the art of selection may, by showing nothing but the truth, produce all the effects of the greatest falsehood."[2]

Macaulay was writing about history, but what he said is applicable when we form any opinion. We can put together nothing but the truth, and the sum of it all is a lie. That may seem surprising, but we can easily see that it is true. Suppose we consider something like nuclear energy, and decide to argue against it. We could say something like this:

> When we operate a nuclear power plant we are using highly radioactive materials, which are extremely dangerous. There are other bad environmental effects; the purification of a ton of uranium produces much more industrial pollution that a ton of coal. Not only that, the amount of energy required to mine and purify a ton of uranium takes a hundred times more energy than to produce a ton of coal.

All of the information in this passage is correct, but by picking out certain information and neglecting the rest, we have reached a false conclusion— we have created a lie, namely, that the use of nuclear energy is both dangerous and inefficient. If, on the other hand, we want to argue in favor of nuclear energy, we can say something like this:

> A ton of uranium in a reactor produces about 10,000 times more energy than a ton of coal. The mining of uranium is much safer than the mining of coal, which is done at the cost of many lives. It is true that one is dealing with radioactive substances, but there is no problem about protecting people against it.

Again, the information is correct, but now we have been led to another falsehood, the conclusion to be drawn being that there are no dangers and huge advantages from the use of nuclear energy. In each case we have deliberately chosen just the information that suits the conclusion we wanted to reach. The truth, of course, falls between the two extremes, and the wise person will come to a more balanced conclusion. There are no simple rules about how this should be done. We should make a sincere study of all of the valid information, give all of it due consideration, and come to a rational and coherent conclusion. We should take particular care to put aside our preconceptions, some of which may be deeply imbedded in our minds.

My favorite example of deliberate misleading involves an enterprising American high school student who composed a factually correct petition that ran somewhat as follows:

> In the vapor state dihydrogen monoxide can cause severe burns. It is an essential factor in the corrosion and erosion of metals and other substances. It is always present in cancerous tumors and in the bodies of persons dying of heart attacks and strokes. It is present in large amounts in

victims of drowning. Since it is colorless, odorless, and tasteless it often escapes detection when present in small quantities.

We recommend that this dangerous chemical should be banned.

He went door to door with it and everyone signed except one person, who apparently knew Latin prefixes well enough to know that dihydrogen monoxide is nothing but H_2O, water. Instead of being a dangerous substance, water is, of course, essential to life.

Propagandists, salesmen, and politicians make a business of selecting information in such a way as to deceive us. It is easy to do. We can select other information to give us a different lie altogether. As Lord Macaulay said, unless we select our information correctly and fairly, we can produce the greatest falsehood, the biggest lie.

Intelligent and fair-minded people try to select information in an unbiased way, using the *judicial* method, the method that is used by a judge and should be used by a jury. When it is used by scientists and other scholars we call it the *academic* method. The essential characteristics of the judicial method are objectivity and coherence; conclusions must be unbiased and fit together logically.

Some people call this the *scientific* method, but I have never liked that expression, because it seems to imply that it is a method used only by scientists. But the unbiased selection of information is important in all intellectual endeavors—in history, literature, economics, political science, journalism, and so forth. It is also used, or should be used, by medical practitioners, detectives, automobile mechanics, and plumbers. As far as possible, it should be used by all of us all of the time.

It is unfortunate that people sometimes use the word *academic* in a derogatory sense, as if the academic method were in some way impractical. The stereotypical academic is supposed to be constantly losing his keys, his address, and sometimes his mind. The truth is that all great advances are made by people who use the academic method.

Although the academic method is the most useful one for most purposes, it is by no means the only one. In public life, an entirely different method, the *adversarial* method, is used very widely by lawyers, politicians, and propagandists. In civil or criminal trials, for example, lawyers present a deliberately biased argument on one side of the case, and other lawyers then present a one-sided argument on the other side. Finally, a judge or jury comes to a decision—using, we hope, the judicial method. The system has obvious weaknesses. The New England poet Robert Frost once said, rather cynically, that a jury was a group of twelve people chosen to decide who has the best lawyers. There are some well-known cases in which the adversarial method has led to the conviction of an innocent person, and cases in which obviously guilty people have been freed on legal technicalities. I do not suggest for a moment that the adversarial method should be discarded for legal cases; it has proved to be the best method on the basis of much experience. However, there must be safeguards to make it as effective as possible. One such safeguard is that the lawyers who argue the case are not the ones who make the final decision. That is done by judges and juries, who are supposed to be impartial and base their decisions on the evidence and arguments presented. When this system works smoothly, as it often does, it is probably the best way to come to a decision on a legal matter.

There are other circumstances in which the adversarial method works well. A member of a debating society will, for example, select information to produce a biased argument in favor of a proposition, and the opposite case is also put forward. The most important condition for the success of the adversarial method is that arguments for both sides should be presented competently at about the same time; otherwise serious errors can result.

The adversarial system is also used in politics, where it creates serious problems. In a modern democratic country decisions are made by democratically elected members of a legislative body such as a congress or parliament. The system has developed, perhaps inevitably, into one in which nearly all members of legislatures belong to well-organized political parties with rather inflexible policies. When debates occur, they are adversarial. Arguments presented by members of a given party will be based on information that favors the views of that party. What makes the system so unreliable and ineffective is that the safeguards are often inadequate. In a trial before a judge, or a judge and jury, the final decision is not made by the people making the biased arguments, but in a legislature the outcome is usually decided in favor of the party having the most members.

There are other serious dangers associated with political processes. In democratic countries it is taken for granted that all citizens exert the same political power. This does not always lead to the most satisfactory outcome. With issues like health care and the economy, is it really sensible to allow decisions to be made by popular vote, when we know that most of the electorate has little understanding of the complex issues involved? Is it satisfactory to elect national leaders by popular vote, or is it not better to let those citizens elect a number of legislators who, on the basis of their experience, will choose the leader of the country? In national elections, is some form of proportional representation not better than the simple system of one person, one vote? Do we really need political parties at all? Many democratically run organizations, such as universities and municipal governments, run well without political parties; why could not national governments do the same? These are surely matters that require careful consideration. The important thing is to try to minimize the fundamental weakness of democracies, which is that they are too much controlled by the opinions of the uninformed or prejudiced.

In a few cases, the use of the adversarial process in politics has even led to disaster. If a political party in power is overwhelmingly strong, the result can be dictatorship. Adolf Hitler came into power without the use of force, and at the beginning he was supported by some people who wanted reform by democratic means. The trouble was that he became so strong, and his opposition so weak, that in the end his lies prevailed completely. Hitler and his cronies became adept at selecting the information that suited their purposes, and suppressing the rest. There is an old story (I think it was in *Punch*) about Dr. Josef Goebbels, the Nazi minister of public enlightenment and propaganda. Someone asked, "Of what is Goebbels a doctor?" Answer: "The truth."

The rise of a dictatorship is an extreme case, but there are many less disastrous but still unfortunate situations that arise from the use of the adversarial system. Often, important political issues are never properly resolved because an entrenched political party is opposed to it on principle. The adversarial method has become so much a part of modern political systems that those involved have become insensitive to its dangers. Politicians who may be intellectually honest in other respects lose their ability to think in an impartial way when they are subjected to the prejudices of their political party. It is probably too much to hope that political parties and the resulting adversarial system will ever be supplanted; that would involve too much drastic reorganization. Perhaps the best we can expect for the next few hundred years is that some of the traditions of modern party politics will be relaxed, so that politicians will be more concerned with the good of society rather than with the success of their party.

The adversarial system is generally used in government commissions, where it also has serious weaknesses. Commissions can drag on in a very inefficient way. Much time would be saved if the adversarial system were replaced by the judicial method, and those asking the questions simply endeavored to ascertain the truth. It is hard to see any difficulty about that; it is what historians, literary scholars, scientists, engineers, detectives, and plumbers do (or should do) all the time.

The adversarial method—the selection of information with the deliberate objective of creating a one-sided argument—is often used by propagandists. It is easy to select information that will support a particular case; one does not have to be particularly clever. In fact, many of the people who do this kind of thing are not clever at all. They have made up their minds without properly considering all the evidence, and then select the evidence that suits their case. There are also people who are paid to present a biased case. This kind of dishonest thinking goes on all the time, and we should do our best to counteract it.

There are many people nowadays who are anti- this and anti- that, and quite often it is perfectly acceptable to be anti- something. Surely we should all be antiwar and anticruelty. But when people say, for example, that they are antinuclear energy, I think we must listen to their arguments in a critical way. Are these people really looking at all of the valid information available, or just what suits their particular case? I have already given examples of antinuclear and pronuclear arguments. It is quite all right for a person to consider all of the information available, weigh it carefully, and come to the conclusion that nuclear energy plants should be outlawed. One can respect such a person, even though one might not agree with the conclusion. What is intolerable, however, is when people deliberately select their information to produce a biased conclusion.

Any argument is greatly weakened when the information is selected in a biased way. I think, for example, that environmentalists would exert much more influence if they employed the judicial rather than the adversarial method. It is undeniable that there must be standards and safeguards to protect the health and welfare of the earth's inhabitants. They should, however, be imposed with all factors taken into account. Sometimes, in expressing and developing their ideas in a biased way, people overlook problems that are more important than the causes they are espousing.

*   *   *

We make proper selections of information by using our intelligence. Intelligence literally means "choose between," from the Latin *inter*, meaning between, and *legere*, meaning "to choose." It is interesting that the word *intelligence* comes from the same Latin verb, *legere*, as the word *select* (*se-legere*), which also means "to choose between." In other words, intelligence and selection mean just about the same thing. Something more than intelligence, or choosing between, is required, however.

We all know intelligent people who sometimes seem to act foolishly, and less intelligent ones who always make the right choices. Obviously we should make our selections by using our intelligence and wisdom to the best of our ability. Whenever the adversarial system is used, in the courts, in politics, or in propaganda, it should always be used intelligently and wisely to avoid error.

If we do not use our intelligence and wisdom constantly throughout our lives, our mental capacity will become atrophied. The main function of education is to sharpen our intelligence, which means that we improve our powers of selection—our intelligence and wisdom. The word education does not mean filling a person with information. The Latin origin of the word *education* tells us that it means a drawing out. The brain is like a muscle: it must be drawn out and constantly exercised to keep it in good condition. I remember hearing of a man—it must have been Oscar Wilde—whose attitude toward physical exercise was that whenever he felt he needed to take exercise he lay down until the feeling passed. There is an old joke about two Mandarin Chinese who watched with amazement two Englishmen playing tennis; why, they wondered, did they not pay a servant to do that? The eminent physical chemist Walther Nernst was annoyed when two brothers who were his research students spent time playing tennis. "Your father is rich," he said to them. "Ask him to send the money to pay someone to knock those silly little balls about."[3]

Today most people seem convinced that exercising their bodies is a good idea. There is not so widespread a fad about exercising our brains, but there should be, since brains soon deteriorate if they are not used.

Above all else, we should all strive to counteract the dangerous tendency toward selecting information in a biased way—thus creating lies from the information that is now so plentiful. In other words, we must rise above mere information, and do our best to make intelligent decisions.

Chapter 2
The Nuts and
Bolts

Science is at no moment quite right, but it is seldom quite wrong, and has,
as a rule, a better chance of being right than the theories of the unscientific. It is, therefore, rational to accept it hypothetically.
—Bertrand Russell, *My Philosophical Development*, 1959

In scientific work we are concerned solely with the processing of a certain
type of information, that which is obtained from the material world by
observation and experiment. We devise theories and hypotheses that provide the best understanding of this type of information and that lead to
predictions testable by further observations and experiments. In the next
few chapters we look at these procedures in several different branches of
science. We will also discuss the main conclusions drawn by scientists
about the composition of matter, the origin and development of the universe, and life on Earth.

The essential feature of science is that attention is focused solely on the
results of observations and experiments. Authoritarian writings, religious
faith and beliefs, metaphysical arguments, nationalism, and emotions are
strictly taboo in scientific work. The rules were clearly stated in 1662 in the
Charter of the Royal Society, the aim of which was to "improve the knowledge of naturall things, and all useful Arts . . . by experiments—(not med-
dling with Divinity, Metaphysics, Moralls, Politicks, Rhetoric, or Logick)."
It is not that the early members of the society were irreligious; on the contrary, the charter includes the statement that their work would "advance the
glory of God, the honour of the King . . . and the general good of

mankind." The attitude at the time was that anything discovered would show how wonderfully well God performed the act of creation. But they established the sensible policy of keeping science and religion in their places, not interacting with one another. It is unfortunate that, even in this so-called enlightened age, there are still people who disbelieve science if it appears to conflict with religious writings.

Scientists are often disconcerted by the strange ideas that people seem to have about science. I recently saw the headline: "Love—Science Says You'll Get over It." I looked with excitement to learn about this important scientific investigation, but was disappointed to find that someone had merely carried out a survey, telephoning people to inquire about their amatory experiences. This kind of thing has nothing to do with science. In much of this book I try to explain what science is really about.

In this chapter we will look at the methods used in physics and chemistry, which are the most basic of the sciences. Physics deals in a general way with the behavior of matter and energy. Chemistry deals more specifically with particular kinds of matter, namely, the elements and their compounds. A scientist working in any field must have a sound understanding of these two sciences, which provide the "nuts and bolts" for work in all of the sciences. Facility with mathematics is also essential to any scientist, since mathematics enters into every branch of science, particularly into physics and chemistry. These two sciences are often called hard sciences.

There is a significant, although not sharp, difference between the so-called hard and soft sciences. The word *hard* is perhaps unfortunate, because it is easily misunderstood. What we do *not* mean is that the science is hard to understand (although some people may think so). Helpful definitions given by the *New Shorter Oxford Dictionary* are "factual, reliable, substantiated; unable to be denied or explained away." For the adjective "soft" the *Oxford Dictionary* gives twenty-three meanings, of which the most relevant for us is "of a science as to method: not amenable to precise mathematical treatment, or to experimental verification or refutation."

What complicates the situation is that almost all of the scientific disciplines have hard and soft areas. By scientific discipline I mean the main divisions of science—physics (including applied mathematics, which is a mathematical formulation of physical problems), chemistry, astronomy, biology, and geology. Of these disciplines, physics is the hardest; in fact, nearly all of it is hard. This is certainly true of the laws of mechanics, of thermodynamics, of electricity, and of the fundamental structures of atoms and molecules. All of these laws can be formulated mathematically, and the conclusions about them have been amply confirmed by experimental data. Here and there we find areas of physics that are less hard, but as time goes on and more experiments are done, the uncertainties become resolved.

As to chemistry, some of it is hard and some of it soft. Until the latter part of the nineteenth century most of chemistry was soft, consisting of a mass of descriptive material about how chemical substances react, with little theory to connect it all together. As time passed, however, and with advances in physics, more and more chemistry became hard. There is one important branch, physical chemistry (or chemical physics, almost the same thing), which is very hard, because it is really physics applied to chemistry; that is, the methods of physics are used, but they are applied to chemical problems.

Physics is entering into chemistry more and more. Until fairly recently, organic chemistry (the chemistry of substances that contain the element carbon) and inorganic chemistry (the chemistry of those that do not) were largely descriptive subjects. Now, physical methods are applied to the structures of compounds and their behavior, and many aspects of these subjects are very hard. A publication by a chemist often reads very much like a paper in physics.

Many people seem to think that scientists proceed by assembling a vast amount of information, then formulating a theory or explanation that best fits the facts. This is the way a detective often solves a crime, or a judge sums up a case in court. It is the way some very routine and unimportant science is carried out. Every research scientist of any competence, however, would agree that this is *not* the way good scientific work is really done. But before discussing the way information is used in the sciences, we should consider the different kinds of information.

We can classify reliable scientific information—a collection of facts or data—as either *specific* or *general*. Examples of a specific or individual fact is that Mount Everest has a particular height, and that a meteorite that has struck the earth has a certain mass. An example of a general fact is the value of the charge on an electron. The information is general because there is no reason to doubt that the same charge is borne by every electron in the universe; we don't have to check each one.

In the hard sciences, general, repeatable, and reproducible facts are most important. Of course, in establishing general facts, such as the charge on an electron, scientists had to take into account many individual facts. In technologies relating to the hard sciences, individual facts are important; a clinical chemist, for example, may specialize in measuring the amounts of chemicals in individual patients, a matter of great importance in therapy. But advances in pure chemistry and physics are concerned with facts that are of widespread significance. We shall later see that specific facts may be more important in the soft sciences like geology and biology than in the hard sciences.

Before going any further, I should again emphasize that much information is simply wrong. Much of what appears on the World Wide Web is incorrect, placed there by persons who have no particular expertise, are often careless about checking their sources of information, and sometimes blindly advocate a particular product or point of view. Books are a more reliable source of information. The publication of a book is an expensive undertaking, and publishers tend to take care over what they publish, emploing reliable people to check what has been written.

In the preceding chapter I defined information as a collection of facts, errors, and meaningless statements. A fact is something that we have no reason to doubt; it remains a fact until shown to be otherwise. In scientific work it is of paramount importance to establish that information is reliable. A necessary condition is that scientific information must be *public*, that is, it must be available to all who want to carry out further investigations, and it must be confirmed by other people. This criterion at once excludes phenomena like miracles, extrasensory perception, spirit reading, prophesying, visitations from outer space, and the Loch Ness Monster; these fail the test of valid scientific information simply by not being public.

Information published in the scientific literature is, on the whole, reliable, because of the checks normally made by journal editors to ensure high quality. Perhaps a more serious problem than incorrectness in the scientific literature is that overzealous editors sometimes suppress papers that are worthy of publication. In some cases the paper is later published, perhaps in another journal, and the erroneous editorial decision becomes exposed.

There are, of course, some exceptions to the general reliability of scientific information that is announced. Incorrect information is occasionally reported as a result of sheer dishonesty, or for personal gain or personal prestige. For example, in the nineteenth century the American swindler Robert Keeley convinced investors that he could create vast amounts of what he actually called atomic energy from a small amount of water. Scientific dishonesty is always found out, sooner rather than later. In Keeley's case the fraud was only discovered after his death, when his demonstration apparatus was found to contain some highly suspicious features.

Fraudulent science for personal prestige, in some cases just to gain a university degree, is by no means uncommon; most university teachers meet a little of it during their careers.

The reporting of incorrect scientific information more often results, not from fraud, but from a genuine mistake. Sometimes the person responsible becomes self-deluded; they repeat a mistake many times even though there may be no conscious intention to deceive. In 1903, shortly after the discovery of various rays, such as x-rays and the rays emitted by radioactive substances, the Frenchman René Prosper Blondlot announced the discovery of a new kind of radiation, which, in honor of his native city Nancy, he called "N-rays." These rays were alleged to increase the brightness of a spark and produce other effects; they were supposed to be different from x-rays and other known rays. Since Blondlot was a respected physicist some credence was given to his finding, especially in France, but other scientists failed to reproduce his work; Blondlot simply said that they lacked the necessary skill. Over a period of several years over one hundred scientists worked on N-rays and over three hundred scientific papers were published about them. The rays were produced by Blondlot by refracting light with an aluminum prism, and they produced faint spots on a screen where Blondlot, but hardly anyone else, could see them. The deathblow to the existence of N-rays was given in a amusing way by the American physicist R. W. Wood. On a visit to Blondlot's laboratory, Wood surreptitiously removed the prism and put it in his pocket, but Blondlot, unaware of this, insisted that he could still see the spots on the screen. That incident convinced everyone but Blondlot, who never conceded that he was mistaken; self-delusion had consumed him.[1]

A similar error was made in connection with what became known as *intensive drying*. The British chemist Harold Baily Dixon (1852–1930) investigated a number of chemical reactions that occur explosively. He found that the reaction between carbon monoxide (CO) and oxygen occurs only if a small amount of water is present; when that is the case, passing a spark through the mixture causes an explosion to occur, with the formation of carbon dioxide (CO_2). If, however, the gases are dried very carefully, no reaction occurs. This seems to be correct as far as that particular reaction is concerned. However, one of Dixon's students, Herbert Brereton Baker (1862–1935), performed experiments on a number of other reactions between gases, and convinced himself—but hardly anyone else—that no such chemical reaction will occur in the complete absence of water. For example, Baker mixed hydrogen and oxygen in a tube together with a powerful drying agent and left the tube alone for many years; finally, he passed a spark through the mixture, and no explosion occurred. Many people tried to repeat this experiment, but found that reaction did occur; Baker said they had not dried the gases long enough—an unanswerable reply. There is a story that on one occasion Baker announced to a distinguished audience that he would demonstrate that a mixture of hydrogen and oxygen that he had been drying for many years would not explode. When the mixture did explode, Baker was quite unmoved, blandly remarking that his drying had obviously been too brief.

In the end everyone became convinced that (except for special cases such as the carbon monoxide oxidation) chemical reactions can occur in the absence of water. Baker was undoubtedly honest in his conclusions; his error probably arose when the materials he used for drying introduced something into the system that inhibited chemical reaction (examples of such inhibitors are well known in chemistry). Much of Baker's other work was good, and he had a distinguished career.

A similar, but much more widely publicized, situation arose in 1989 about *cold fusion*. In the spring of that year Stanley Pons and Martin Fleis-chmann of the University of Utah announced that they could produce large amounts of heat by the expenditure of tiny amounts of electrical energy. The immediate reaction of much of the scientific community was that the results were unbelievable; such large amounts of energy could only be pro-duced if nuclear processes were occurring, and, under the conditions of the experiments, that possibility seemed remote. Largely because of the some-what sensational way in which the results were announced, a considerable stir was produced. There was some confusion about the claim for a few years; some said that they had confirmed the findings, while others said that they could not do so. Complications arose because it was difficult for people to be sure that they were repeating the experiments under the orig-inal conditions; also, there is evidence that something rather odd and unexpected does occur in the reaction, but not what Pons and Fleischmann had asserted. In the end, conclusive experiments by many other scientists showed that the original claim could not be sustained.[2]

These examples are instructive, because they show that things are not necessarily cut and dried even in hard science. There can be uncertainty and controversy, but it is usually resolved fairly quickly.

There is no clear pattern of scientific discovery, or of technical advance. Some writers on this topic, even scientists themselves, tend to oversimplify this matter. Active research scientists often read discussions of the "scien-tific method" written by nonscientists, but find that they cannot relate it to the way they themselves work.

Earlier, I explained how the term "judicial method" or "academic method" is preferable to "scientific method." It took many centuries for the method to evolve. It was first used, though not entirely satisfactorily, by the Greeks a few centuries BCE. Aristotle, for example, made the serious error of allowing intuitive ideas, unsupported by experiment, to intrude into his science. One of his principles was that circular motion is the "natural" type of motion, and he explained the movement of the planets in this way. Another of his principles was that every object has a natural tendency to reach its "natural" place, the center of the universe, which to Aristotle was the center of the earth. Also, Aristotle was convinced that women have two fewer teeth than men. He married twice, and could easily have made an observational test, but apparently did not bother to do so.

The reasons for the intellectually Dark Ages that intervened between the Greek civilization and the Renaissance has been much debated by histo-rians. During the Dark Ages, there was a general conviction that the truth about the world and its origins could be derived from studies of religious writings, such as the Old Testament, and from mystical and astrological practices. Theology and astrology are rather unexacting intellectual activi-ties, demanding a lot of plodding but not much originality, imagination, or

precise thinking. It is not surprising that they were favored over the intellectually more demanding scientific investigations that were to come later.
Eventually, theological, mystical, and astrological studies were shown by the
advance of science to throw no light whatever on the workings of nature.

The judicial or academic methods were reinvented in western Europe
during the Renaissance. The person who played the dominant role in this
reinvention was the statesman and philosopher Francis Bacon
(1561–1626; see figure 1)—the same man who some believe to have
written the plays usually attributed to Shakespeare. His ideas about philosophy and science were set out in his *Novum Organum* (1620) and *New
Atlantis* (1627). These writings emphasized that all theories in philosophy
and science must be based on firmly established information, acquired by
careful observation and experimentation; there is no place for ideas based
on religion or metaphysics. Bacon emphasized that intuitive principles—
like those used by Aristotle—must not be taken for granted, but tested
against the experimental evidence. This is the approach made by scientists
today, and it is sometimes referred to as *positivism*.

One way in which progress is sometimes made in the hard sciences is
by the formulation of what are called *empirical laws*. An empirical law
simply groups together a number of related facts and summarizes them,
often in the form of a mathematical equation. For example, suppose that a
gas such as air is contained in a cylinder with a piston, and has a certain
volume and pressure. We find that if we double the pressure the volume is
reduced to almost exactly one-half. The results can all be summarized in a
very simple equation,

$$\text{pressure} \times \text{volume} = \text{constant}$$

where the constant is always the same for a given amount of a particular
gas at a specified temperature. This equation, usually known as Boyle's law,
illustrates a *reciprocal* relationship between pressure and volume: when one
of them goes up, the other goes down by the same factor.

Another important empirical law relating to gases is Gay-Lussac's law
(often called Charles's law in English-speaking countries). To understand
this law, it is best to use the Kelvin or absolute scale of temperatures. The
Celsius and Kelvin scales, closely tied to one another, are universally used
by scientists, since they greatly simplify scientific relationships. The Celsius
scale is very close to the old Centigrade scale, which was based on 0 °C for
the melting point of ice at normal pressure, and 100 °C for the normal
boiling point of water. (The Fahrenheit scale is based on 32 °F and 212 °F
for these fixed points.) The value of the Kelvin temperature is the value of
the Celsius temperature plus 273.15. For example, if the temperature is 25
°C, which is 77 °F, the absolute temperature is 298.15 K.

Fig. 1. Francis Bacon, English statesman and natural philosopher, suggested the methods used by scientists and other scholars today. He held high office under King James I of England, becoming Lord High Chancellor in 1618. Convicted of accepting bribes, he was banished from court and office in 1621, after which he devoted himself to contemplating the methods of science and to writing several important books and articles. (Courtesy of the Horace Howard Furness Collection, Annenberg Rare Book and Manuscript Library, University of Pennsylvania.)

Gay-Lussac's law states that, at a given pressure, the volume of a gas is proportional to the absolute or Kelvin temperature. Gay-Lussac's law can thus be written as:

$$\text{volume} = \text{constant} \times \text{absolute temperature}$$

Many other experimental results in the hard sciences have been summarized by empirical laws. Another example is Ohm's law. If a voltage is applied to the ends of a wire, the strength of the electric current passing through the wire is doubled if the voltage is doubled. In mathematical language we say that the current is proportional to the voltage and we can write the equation:

$$\text{electric current} = \text{constant} \times \text{voltage}$$

This relationship was first pointed out in 1827 by the German physicist Georg Simon Ohm (1789–1854). The constant that appears in the equation is known as the *conductance*; the greater the conductance of a wire, the greater the current running through it when a given voltage is applied. The reciprocal of the conductance (i.e., one divided by the conductance) is called the *resistance*; the greater the resistance of a wire, the smaller the current when a given voltage is applied. We can write Ohm's law as

$$\text{resistance} = \frac{\text{voltage}}{\text{current}}$$

Empirical laws are common in all of the sciences. They may not be accurate under all circumstances, but they are nevertheless useful as guides in many applications. Representing no more than a first step in the understanding of experimental information, they do not provide any deep understanding of the facts, but merely summarize them in a convenient way.

The collection of data and the formulation of empirical laws do not take us very far. It is also important to interpret data in terms of a theory or a fundamental scientific law. Some explanation is necessary here, since, unfortunately, scientists tend to use the words *law* and *theory* rather loosely. This is partly for historical reasons; scientific laws and theories tend to change their status as more investigations are made. Scientists usually understand one another when they refer to particular laws and theories, but nonscientists may easily be confused.

In addition to empirical laws, such as described above, there are also some scientific laws used in science that have a much more fundamental significance. For example, there are certain laws that are called *universal* laws since, as far as we know, they apply throughout the universe. One of the universal laws is the law of gravity, discovered by Sir Isaac Newton (1642–1727; see figure 2). Newton also formulated three laws of motion, and Einstein's law of relativity is another universal law. Scientists usually still refer to the "theory" of relativity, but, now that it has been so well confirmed experimentally, we take it for granted that it has the same basic significance as Newton's laws.

Two other universal laws are also usually referred to as theories—the atomic theory and the quantum theory. According to the atomic theory, matter is made up of particles called atoms. According to the quantum theory, energy comes in packets, as does light. When both of these ideas were first proposed they were indeed theories, but they have now been confirmed so many times that they have the status of universal laws.

The empirical laws we discussed earlier do not have the same status as the universal laws. They are *derivable* laws. This means that, starting with the particles that exist in nature, we can work out mathematical treatments of the laws based on the universal laws. For example, Boyle's law, which relates the pressure of a gas to its volume, can be derived from the kinetic theory of gases. This mathematical treatment recognizes that the pressure of a gas is due to the bombardment of molecules on the walls of the vessel in which the gas is maintained; it leads to the empirical law by applying Newton's universal laws of mechanics to the constituent particle (the molecules) of the gas.

The word *theory*, as used in science, often gives rise to confusion. In ordinary life the word is often used for a rather trifling idea. Someone is said to have a "pet theory" about something, which usually means that we should not take it seriously. Often people say that some idea is "just a theory," a rather vague explanation. The expression "just a theory" is not appropriately applied to the quantum theory or the theory of evolution. These are not "just theories"; they have been confirmed by so much experimental and observational evidence that there can now be no doubt about their validity. It would be perfectly correct to call them laws, but when they were first formulated they were just theories, not accepted by everyone, and the name "theory" has stuck. In science, then, a theory is an important idea whose validity is transformed by experiment or meticulous observation into certainty, thus becoming a scientific law.

Scientists also often use the word *hypothesis* to mean a plausible idea that is put forward. It is a tentative theory, an idea that later may become a theory when further information has accumulated. We may legitimately call a hypotheses "just a theory." People are sometimes confused about Isaac Newton's attitude toward hypotheses, and quote out of context his phrase "hypotheses non fingo," "I make no hypotheses." From this some

Fig. 2. Isaac Newton. While still a student he developed a theory of gravitation and propounded laws of motion of the planets. He was the first to split a beam of light into its spectrum and then recombine the colors into white light. He also discovered the binomial theorem and a system of calculus.

people have mistakenly concluded that Newton never made hypotheses. However, he made his comment with regard to a particular problem, gravity, and he meant that he made no hypotheses about the ultimate cause of gravity, which he admitted he did not understand. In the rest of his work he certainly did propose many hypotheses and theories. He frequently used the words *hypothesis* and *theory* in his writings, and several times in the titles of his articles.

The interrelationship between scientific information, laws, and theories is not entirely simple. Today hardly any observational or experimental work is done to test the universal laws of nature, since they are already so well established. Also, competent scientific work is never done in isolation, with no purpose in mind. The facts are always obtained in relation to empirical laws, and often with the intention of establishing or testing theories. In other words, many facts are *theory-laden*, in that their discovery, and sometimes their interpretation, depends on the scientific laws and theories.

For example, after the generalization known as Boyle's law became known, most of the observations on the pressure and volume of gases were made with the law in mind. An interesting point arises here, relating to the reliability of data obtained in this way. Scientists may be tempted to adjust data a little so that they will fit a law. It has even been suggested that Newton "fudged" some of his data on the movement of planets so that they would fit more accurately the law of gravitation that he had propounded. We must recognize that important advances are often made when scientists discover deviations from empirical laws. For example, deviations from Boyle's law led to the realization that the molecules of a gas attract one another.

Measurements of applied voltage and current are now always made with reference to Ohm's law, and again there may be some bias in favor of giving support to the law. Important advances have often been made by people who have discovered unexpected behavior. For example, certain substances, known as superconductors, have been found to have an abnormally high electrical conductivity at low temperatures. Work on the properties of such substances has revolutionized the electronics industry.

Some scientific information is heavily theory-laden. For example, scientists accept the fact that the universe was formed several billion years ago, even though this information is obviously not obtained by any direct measurement. Instead, the fact is deduced from a complicated network of theory, which in turn is based on extensive astronomical observations. In spite of being so indirect, the conclusion about the age of the universe is convincing to all who have made a study of the evidence; it is reasonable to refer to it as correct information, even though the exact age of the universe cannot be given precisely. The same is true of the theory of evolution.

When it was first put forward, it was "just a theory," but it has led to such a vast amount of supporting data that evolution is now regarded as a fact.

The dividing line between fact and theory in science is by no means a sharp one. No competent scientist carries out experiments without any theory in mind. There is a variation in the reliability of information, even in the hard sciences. Nevertheless, the fundamental structure of the sciences is sound, since the theories that have been developed have been extensively tested.

A scientific theory is an explanation of how and why things happen as they do, with the emphasis on *how*. A theory leads to predictions about what should happen if certain other experiments or observations are made.

How scientific theories originate is a matter of some importance, and is often misunderstood and oversimplified. I emphasized earlier that it is wrong to think a research scientist proceeds by assembling data and then formulating a theory to explain the data. As a scientist or engineer struggles through the vast amount of reliable experimental information available, there is no well-marked path to a satisfactory theory. There are not even any clear-cut rules of the road, but only a few general guidelines. A successful scientist feels free to use any methods, and must not be afraid to use imagination and intuition, and to break rules. When the great chemist Jacobus Hendricus van't Hoff (1852–1911) delivered his inaugural lecture to the University of Amsterdam in 1878, he titled it "Imagination in Science." He said he had made a special study of the way in which the great scientific advances of the past had been made, and stressed that imagination and observation were both of great importance. As the distinguished American physicist Percy Williams Bridgman (1882–1961) put it pithily: "Use your noodle, and no holds barred."

This may appear surprising; surely there are some things that one must not do, such as invent data. It would indeed be wrong to do so, but sometimes one does better by ignoring some of the data. This point, which some people may find disturbing, requires a little discussion.

I mentioned earlier that Newton is believed to have manipulated his data a little. He was the first to formulate a universal law of gravitation and to apply it to calculating the orbits of planets around the sun. In doing so he sometimes found that there were discrepancies between his calculations and the observed data. Newton thought nothing of brushing under the carpet any data that did not agree with his calculations, and for this he has sometimes been criticized. I do not really think that this is fair. We now know that some of the discrepancies were due to the attractions of planets not then discovered. As it has turned out, Newton's theory was essentially correct (aside from Einstein's relativity corrections, which do not come into the present discussion). It would have been unfortunate if Newton had failed to put forward his great theory because of the small discrepancies that he noticed.

Incidentally, Newton's whole scientific career is an object lesson on the importance of not sticking to a so-called scientific method. In many ways Newton's methods were quite unscientific! Besides keeping quiet about facts that did not fit his theories, he was much guided by metaphysical and religious arguments, which any modern scientist would avoid. In spite of all these weaknesses, he surely was one of the greatest scientists who ever lived, because of his insight and imagination.

In formulating theories, scientists usually apply a principle, suggested in the fourteenth century, known as Occam's Razor. It states that *"Entia non sunt multiplicanda praeter necessitatem"* or "entities are not to be multiplied unnecessarily." As modified for scientific purposes, the principle is that explanations should be as simple as possible, with no unnecessary frills. In science it is always best to strive for simplicity, since unnecessary details to a theory may be proven wrong, even though the basic theory was essentially correct. Of course, theories are often expanded as additional information is obtained. William of Occam or Ockham (c.1285–c.1349), the originator of the principle, was a Franciscan philosopher and theologian who proposed his principle to apply to theological doctrines, which he thought were becoming unnecessarily complicated.

Occasionally, facts that are not easy to obtain directly are deduced from theory. A good example is provided by the work done in the 1840s on Saturn's rings by the great Scottish physicist James Clerk Maxwell (1831–1879; see figure 3). At the time, astronomers had observed three concentric rings about Saturn, all in the same plane. They knew that at least some regions of the rings must be quite thin, since in some areas the planet behind could be plainly seen. Aside from that, their structure was an intriguing mystery; they might be solids or they might be fluids (liquids or gases). Maxwell carried out a careful theoretical and highly mathematical treatment, and concluded that the rings could not be solid or liquid, since the mechanical forces acting upon rings of such immense size would break them up. He suggested instead that the rings must be composed of a vast number of individual solid particles rotating in separate concentric orbits at different speeds. He even drew some conclusions about how large the particles would be and how fast they would move.

For many decades there was no way of testing Maxwell's conclusion, but in the latter years of the twentieth century, observations—particularly those from *Voyager* spacecraft—confirmed Maxwell's conclusions. The particles are composed of impure ice, or at least are ice-covered. Radar observations have even confirmed the range of masses and speeds predicted by Maxwell. This was a remarkable achievement on Maxwell's part. On the basis of theory he arrived at factual information that could not be directly obtained for almost a century and a half.

I mentioned earlier that scientific work is best done with a theory in

Fig. 3. James Clerk Maxwell made pioneering contributions in several fields. He is particularly noted for his treatment of the distribution of molecular speeds in a gas, which led to the development of important statistical methods in science. He is also celebrated for his theory of electromagnetic radiation.

mind. It does not matter if the theory is proved wrong; it will nevertheless guide the initial experiments, which may lead to the correct theory. Paradoxically, people occasionally do much better with the wrong theory than with the right one, and there is one rather striking example of this. By the year 1900 physicists had become aware of the properties of radio waves, which they knew to be of the same character as light but with much longer wavelengths. Experiments had shown that radio waves (unlike light) would pass through walls because of the long wavelengths, and that like light they travel in straight lines.

One of the people who was working on radio waves at the time was the Italian Guglielmo Marconi (1874–1937). Unlike almost everyone else working on them, he knew hardly any physics. (Hence, the award to him of the 1909 Nobel Prize for physics amazed and horrified most physicists.) He did not really understand the properties of electromagnetic waves, and did not even realize that radio waves were of that type. He said that his waves were Marconiwaves, and that, instead of traveling in straight lines, they could be controlled by him to go to any point he chose.

This was nonsense, but his ignorance of physics led him to try to send a signal across the Atlantic Ocean. No other physicist thought this was a reasonable thing to try—how could a wave, which has to travel in a straight line, possibly travel so far over the curved surface of the Atlantic? On December 12, 1901, Marconi's assistant sent signals across the Atlantic Ocean, from Cornwall, England, to St. John's, Newfoundland, where Marconi was apparently just able to receive them.[3]

Physicists were quite correct in thinking that radio waves go in straight lines. How, then, can a radio signal go across the Atlantic, in view of the curved surface of the earth? The most plausible answer is that they must be reflected in some way. In 1902 the British physicist Oliver Heaviside (1850–1925) and the American physicist Arthur Edwin Kennelly (1861–1939) postulated the existence of a layer in the upper atmosphere in which they thought that the molecules would be *ionized*, meaning they would carry electric charges. Such a layer would reflect the waves from the upper atmosphere and allow them to travel great distances over the earth's surface. This is another good example of the deduction of a fact—the existence of the ionized layer—from a theory. For some years this idea of an *ionosphere* was "just a theory," but later work has proven its existence.

There are not many instances where great progress in science has been made as a result of someone's ignorance. Almost inevitably it is the best-informed people who produce the most useful experimental results and theories.

From centuries of scientific work, there have now accumulated many important scientific generalizations about which there is no controversy.

This knowledge is based on a vast amount of circumstantial evidence coming from sources of entirely different kinds. With the public, particularly in connection with jury trials, circumstantial evidence seems to have a bad name. However, extensive circumstantial evidence can be overwhelmingly convincing. A particularly lucid discussion of circumstantial evidence was given by Lemuel Shaw, chief justice of the Commonwealth of Massachusetts, in his summing-up at the 1850 murder trial of Harvard professor John Webster. Justice Shaw pointed out that direct evidence can be easily tainted, perhaps by perjury or by an honest mistake. Circumstantial evidence, on the other hand, can be much more reliable, especially if there is the accumulation of such evidence from a variety of sources. In the Webster case, there was a vast amount of circumstantial evidence, all mutually self-consistent, and there was no doubt of his guilt.[4]

So it is in science. A powerful body of direct evidence sometimes appears to lead inevitably to one theory, but it may later turn out that an entirely different theory is preferable. In another situation, however, several completely independent lines of evidence may point to a particular theory, and even though the evidence may not be direct, the case for the theory may be completely convincing. Even in the hard sciences many important theories were initially deduced from circumstantial evidence. At the beginning of the twentieth century hardly anyone doubted the reality of atoms, though their existence had been deduced on the basis of quite indirect evidence; today we have a good deal of direct evidence. All the major concepts of science accumulated a vast amount of circumstantial evidence, often from widely different branches of science. The laws of thermodynamics, the theory of evolution, quantum theory, the theory of relativity, chaos theory, and modern scientific theories of the age of the universe and of life on Earth, are so strongly supported by such evidence that no unbiased person who has studied it can doubt that these laws and theories are close to the truth.

Scientists are apt to say that such theories represent the "truth." They do not mean any absolute truth, but simply that they have found that, in practice, the theories always work, and are useful in providing a framework that will be a completely reliable guide for their future investigations. They mean that they are convinced that theories like the ones just mentioned can hardly be untrue. This point is often misunderstood. People may think, for example, that if they can find facts that are incompatible with what Darwin said about his theory of evolution, they have demolished the whole theory. The theory, however, is so well built that it takes more than a few apparently inconsistent facts to destroy it, just as a well-built bridge cannot be demolished by the removal of a few minor struts.

*   *   *

There is much testimony in the writings of great and creative scientists that the invention of a theory is almost always an act of imagination. There may be flashes of inspiration that have little relation to any particular experimental data, and sometimes they come from subconscious thought rather than from analysis of information. Let us consider a few examples of how some important scientific principles came about.

One of the most important principles in physics in the nineteenth century was concerned with the nature of heat and the relationship between heat and mechanical work. The theory is known as the *principle of conservation of energy*, which means much the same thing as the *first law of thermodynamics*. This law says that energy cannot be created from nothing, and cannot be destroyed; we are all conscious of this today, but it was by no means obvious at first. Many of the ideas about energy were inspired by the functioning of steam engines, particularly by the innovations made by Scottish inventor James Watt (1736–1819). As we will see, energy is an important concept for understanding the world around us. It is therefore surprising that people only began to think about energy in the nineteenth century, and that only after a number of decades was it realized that the total energy of a system always remains constant.

Several people were involved in discovering this law, and they worked in quite different ways, all of them using their imagination and intuition. One of them was the American-born Benjamin Thompson (1753–1814), who had a remarkable career in Europe and later became Count Rumford. He was impressed by the heat produced in the boring of cannon, and concluded that it must be produced by the work done in the process. Another was the German physician Julius Robert Meyer (1814–1878); he did hardly any experiments, but pondered on digestive processes and the work done by humans. The most detailed experiments that proved energy to be conserved were done by James Prescott Joule (1818–1899), an English amateur scientist. Beginning in about 1837, he carried out a variety of careful experiments, and his investigations completely transformed the subject.

Some of Joule's work was on electric batteries, which had been invented in the early years of the nineteenth century. Some people had been led to the idea that there was no limit to the amount of energy that could be obtained if a battery was connected to an electric motor. Steam engines had to be supplied with fuel obtained from underground, and it was realized, even in the early nineteenth century, that the supply of such fuel would eventually run out. Electric batteries, on the other hand, led to what has been called an "electrical euphoria," because it was thought that batteries could produce enormous amounts of energy at no cost. An enthusiastic proponent of this point of view was Moritz Hermann von Jacobi (1801–1874), who constructed what was perhaps the first electric motor. In 1835 Jacobi published a paper that created something of a sensation. He argued that if certain imperfections of the electric motor, such as friction, could be eliminated, a motor would go on accelerating indefinitely, producing enormous amounts of energy. These arguments, although wrong, seemed compelling at the time, and many electric motors were built in the next few years to exploit the idea.

Joule also thought at first that electric motors might be able to produce unlimited amounts of energy. He carried out careful experiments on the mechanical effect that could be obtained from a motor, and related it to the amounts of metal used up in the battery operating the motor. He found that the consumption of a given amount of zinc in a battery would lead to the production of only about one-fifth of the mechanical work that a steam engine would produce from the same weight of coal. Since zinc was also much more expensive than coal, this meant that an electric motor was far from being a competitor to a steam engine for the primary production of energy. For Joule, the "electrical euphoria" was over.

Joule then studied the heat produced by an electric current, and found that it was equivalent to the energy released by the chemical action occurring in the battery. This conclusion was important in showing that, contrary to von Jacobi's prediction, energy could not be created from nothing. With regard to practical use, Joule wrote that "electricity is a grand agent for carrying, arranging and converting chemical heat." This was a sound prediction: in our cars, cell phones, and laptop computers we always carry a convenient source of energy in the form of a battery. Joule later carried out many other experiments that proved that energy is conserved.

This illustrates the way an important scientific conclusion is often reached. It was not a simple matter of looking at experimental data and drawing logical conclusions. Rumford, Meyer, and Joule worked in entirely different ways. A wrong idea—that there was no limit to the amount of work that could be produced by a battery—had been put forward, Joule set out to test the idea, and came to the correct conclusion. We cannot get mechanical work for nothing—this is the lesson of the first law of thermodynamics. From time to time people try to patent so-called perpetual motion machines in which the first law is violated. These are known as *perpetual motion machines of the first kind*, and needless to say, they never work.

The way in which the *second law of thermodynamics* was arrived at is also instructive. It, too, was reached rather gradually, and resulted from the realization that the application of the first law to steam engines was not entirely straightforward. The distinguished physicist William Thomson (1824–1907), now usually remembered as Lord Kelvin, was slow to accept even the first law, although he was on friendly terms with Joule and followed his work closely. Paradoxically, the reason for his initial reluctance was that he had studied heat conversion even more carefully than Joule had. Kelvin knew that if there was an interconversion of heat and work, the process is rather peculiar. Work could be converted into heat without any apparent complications, but not all of the heat produced by a fuel can be converted into work. After thinking about this problem for some time, Kelvin was finally led to a more profound understanding of the restrictions governing the conversion of heat into work.

Kelvin was greatly helped by the ideas of a remarkable young French military engineer, Nicolas Leonard Sadi Carnot (1796–1832), who had made a detailed study of the working of steam engines. He realized that not all of the heat provided by the fuel can be converted into mechanical work; some of the heat must always go to waste. Kelvin followed up Carnot's ideas, and referred to this wastage of heat as the *dissipation of energy*. Like Carnot, he realized that an engine cannot operate if everything is at the same temperature. For example, a ship cannot propel itself by abstracting heat from the surrounding water; the heat must be obtained from something at a higher temperature, and there must be dissipation of heat, some heat passing from a higher to a lower temperature. This is one way of expressing the second law of thermodynamics.

A machine that, although consistent with the first law, is supposed to violate the second law, is referred to as a *perpetual motion machine of the second kind*. An example would be a ship that operated by withdrawing heat from the surrounding waters. It is interesting that a ship designed to do something like that actually put to sea. It was designed by the Swedish-born inventor John Ericcson (1803–1899), who settled in the United States. He did some excellent work, designing the first warship with a screw propeller and with engines below the waterline. Later he built a vessel named the *Ericcson*, which was fitted with "caloric" engines, which were supposed to use the same heat over and over again. Ericcson went considerably too far, since if his engine had worked it would have violated both the first and the second laws of thermodynamics. Instead of crossing the Atlantic, as he had hoped, it had to be refitted with conventional steam engines, but even then the ill-fated vessel sank to the bottom of the sea in 1854. We should not blame Ericcson too much for his mistakes, since hardly anyone understood the two laws of thermodynamics at that time. No doubt his spectacular failure led to a deeper appreciation of that branch of science.

A somewhat different way of expressing the second law of thermodynamics was suggested by the German physicist Rudolf Julian Emmanuel Clausius (1822–1888). He suggested a new physical property, which in 1865 he called *entropy*, and expressed the law by saying that the entropy of the universe always tends toward a maximum value. It would take too much mathematics to explain exactly what entropy is, but it is easy to get the general idea. Entropy is a rather precise measure that relates to the probability of the existence of some particular system. What happens when we shuffle a deck of cards helps to explain the idea of entropy. When we buy a deck, the cards are arranged in a particular way, which we call ordered. It can then be shuffled, and the shuffled deck is more probable than the ordered one, because there are many different shuffled decks but only one ordered one. The shuffled deck has a higher entropy than the ordered deck. Shuffling an ordered deck will almost certainly produce a disordered deck, with an increase in entropy. It is highly unlikely that a shuffled deck will become ordered if it is further shuffled; that would involve a reduction in entropy. Entropy has often been referred to as the "arrow of time," meaning that time cannot go backward but only forward. This is because the state of the universe has a greater probability at a later time than at an earlier time.

We are all familiar with processes that occur naturally and as a result of this tendency for an ordered state to become a disordered one. In all cases there is an increase in entropy. When a lump of sugar is dissolved in coffee, we know that the molecules of the sugar spread themselves throughout the liquid; however long we wait, we do not find the cube reforming itself—although if we could wait a very long time (perhaps longer than the age of the universe) it would do so, and at once dissolve again. If a bottle of perfume is left open, the perfume spreads around the room, and in our lifetimes the molecules of perfume will almost certainly not go back again into the bottle. There is a demonstration experiment in which oxygen and hydrogen gases are brought together, and a flame is put to the mixture; the gases explode with the formation of water (H_2O). However long we wait, a glass of water will not suddenly decompose into hydrogen and oxygen. The reason is that there is a great increase of entropy when the gases are exploded together, largely because heat is given off and is dissipated into the surroundings. In principle, heat given off could assemble in a glass of water and decompose it into hydrogen and oxygen, but the probability of this happening is extremely remote.

Entropy is so subtle a property that many scientists were unable to understand it when Clausius first suggested it. Kelvin, for example, never appreciated entropy, and maintained that the second law can be more easily understood in terms of the dissipation of heat, which is easily visualized. Kelvin's philosophy of science was that everything must be explained in terms of a mechanical model, and entropy cannot be explained this way. Properties like volume, pressure, and temperature can be measured with simple instruments and can be understood even by people who do not know much science. Entropy, on the other hand, is elusive; no instrument can directly measure an entropy change, which has to be calculated in a rather complicated way from data involving heat and temperature changes.

The first two laws of thermodynamics are accepted today without question. They are usually called laws, and they are empirical laws, since they were arrived at on the basis of experimental results. Our understanding of them, however, involves much more than a convenient way of summarizing data, in that they have a strong theoretical infrastructure. The first law is a consequence of the fact that heat is a form of motion; this fact, together with the laws of motion, lead logically to the first law, that energy is conserved. The status of the second law is quite different in that it is a statistical law. The universal laws do not require the second law to be obeyed, and it could in principle be violated. It is true only as a matter of chance, but for systems containing large numbers of molecules, the probability of its being violated is exceedingly small.

Until the middle of the nineteenth century, most scientific theories had been based on models or images that could be clearly visualized. The idea of doing this originated with the philosopher René Descartes (1596–1650). To him and others at the time, it was difficult to understand how gravity could act through empty space. To get around the difficulty Descartes suggested that what seemed to be empty space was filled with something called the *ether*. He ascribed some specific properties of the ether, and regarded it as composed of swirling particles, referred to as vortices. When it was later realized that light also appears to travel through empty space, it was supposed that its transmission also involves some kind of swirling processes occurring in the ether.

Kelvin always insisted on a mechanical model: "I never satisfy myself unless I can make a mechanical model of a thing. If I can make a mechanical model I can understand it." This was unfortunate for him, because during his lifetime two important theories were put forward which were not based on mechanical models, and Kelvin was never able to understand them. One of these was the electromagnetic theory of radiation, which made it unnecessary to think in terms of the ether. The other was the idea of entropy.

A model to explain the idea behind entropy was proposed by Kelvin's friend Clerk Maxwell. The way in which his idea came about is rather unusual and interesting. Scientific ideas usually come to light in formal scientific papers, but Maxwell's proposal was an exception. It took the form of an imaginary supernatural being, later called *Maxwell's demon*, which was born in a letter that Maxwell wrote to his friend Peter Guthrie Tait on December 11, 1867. The point of this letter was to show how, in principle (but hardly ever in practice), the second law could be violated.

Maxwell considered a vessel divided into two compartments A and B, separated by a partition which had a hole in it that could be opened or closed by "a slide without mass" (see figure 4). The gas in A was at a higher temperature than the gas in B. By this time Maxwell was well aware that, in a gas at a given temperature, the molecules will be moving with a variety of

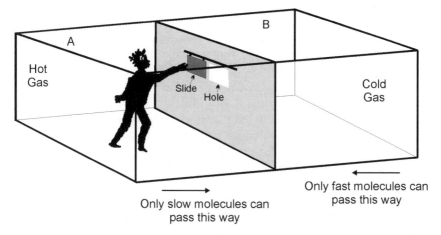

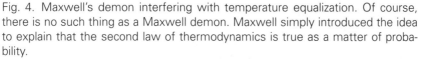

Fig. 4. Maxwell's demon interfering with temperature equalization. Of course, there is no such thing as a Maxwell demon. Maxwell simply introduced the idea to explain that the second law of thermodynamics is true as a matter of probability.

speeds; some will be moving fast, others slowly. Maxwell imagined "a finite being," later called a demon, who knew the speeds of all the molecules. This creature would open the hole for an approaching molecule in A when its speed was low, and would allow a molecule from B to pass through the hole into A only when it was moving fast. As a result of this process, said Maxwell, "the hot system has got hotter and the cold colder and yet no work has been done." Of course, Maxwell did not imagine that his "finite being" could exist; he emphasized that his intention in inventing it had simply been to provide us with an understanding of why the second law of thermodynamics applies; it is just a matter of probability. In a letter to J. W. Strutt (later Lord Rayleigh), written in December 1870, Maxwell commented: "The 2nd law of thermodynamics has the same degree of truth as the statement that if you throw a tumblerful of water into the sea, you cannot get the same tumblerful out again [i.e, exactly the same molecules as before]."

Maxwell's idea of a doorkeeper who could bring about a violation of the second law was soon taken up by others. It was Kelvin who first referred to the creature as a demon, and he later playfully endowed it "with arms and legs—two hands and ten fingers suffice." He had no difficulty in understanding Maxwell's demon, and he understood how it explained the second law of thermodynamics. His difficulty was that he could not appreciate the concept of entropy, since he could not visualize such a property.

Kelvin also failed to appreciate the electromagnetic theory of radiation, put forward by Maxwell. We will defer explanation of this theory until the next chapter, and here just mention one feature of it. One of the first things that Maxwell did was to devise a model for the ether, which was supposed to carry gravitational and electrical forces. According to Maxwell's original model this ether had a rather complicated structure, consisting of spinning material (vortices), some of them electrical and some magnetic. Today this model is only of historical interest, in showing how Maxwell's ideas developed. Over the years he pondered deeply over these matters, and he published an important paper in which he took an entirely different point of view. In this paper he ignored the rather elaborate and artificial model he had proposed for the ether, with its spinning vortices, and concentrated on the propagation of electromagnetic waves through space. The position he took, which is accepted today, is that the mathematical treatment remains valid whatever the nature of the medium through which the waves travel. It is thus unnecessary to postulate the existence of an ether.

Maxwell made an important break with scientific tradition, in that his theory was not based on a model that could be clearly visualized. His famous book *Treatise on Electricity and Magnetism* was published in 1873. A significant feature of the book is that the word ether is mentioned only once. This does not mean that Maxwell had abandoned his belief in the existence of an ether; in his article on "Ether" in the famous 9th edition of the *Encyclopaedia Britannica* (1875), he expressed very clearly his belief in the existence of an ether. He made it clear, however, that his theory of electromagnetic radiation was valid whether or not the ether exists, or whatever its nature.

Since Kelvin required a model to understand a theory, he found himself quite incapable of understanding either Clausius's theory of entropy or Maxwell's theory of radiation. This is somewhat surprising, since he had made important contributions to understanding both radiation and the second law of thermodynamics. His understanding of them was, however, limited by his need for a model, and he was never able to overcome his difficulties and adapt to the new types of theory. Physicists today have become used to theories that cannot be visualized.

As science progresses there is a steady refinement of theories. Historians and philosophers of science sometimes talk of scientific revolutions in which one set of theories is overthrown and replaced by another—or, as they sometimes say rather portentously, one paradigm is replaced by another. Many scientists consider that this is rather overstating the case.

It is sometimes said, for example, that Einstein's theory of relativity rendered Newton's mechanical theory obsolete, but such a statement is misleading. In most of their work, scientists still use Newtonian mechanics and ignore the theory of relativity. Engineers, even rocket engineers, always use Newton's theory, never Einstein's. Deviations from Newton's mechanics only occur when we are dealing with speeds approaching the speed of light, which happens only in special and rather unusual circumstances. Relativity theory therefore does not appreciably affect most of what occurs, but it has had a great effect on the way scientists think about space and time. Similarly, the quantum theory was revolutionary as far as some scientific and engineering work is concerned; in the development of radio, television, and computers, quantum effects certainly have to be taken into account. However, for many practical applications, such as driving a car, it has no noticeable effect.

It usually takes some years for any scientific innovation to be accepted by the scientific world. All people have a natural conservatism, and scientists are no exception. Max Planck once remarked that a new theory becomes universally accepted only after the older scientists have died off; as one wag put it, "science advances funeral by funeral." No doubt Planck was thinking of the long delay in the acceptance of his quantum theory. There is a certain unconscious irony in his remark, since, although he arrived at the theory in 1900, he did not really believe it for a few years. After 1905 he became gradually convinced by Einstein, who somewhat extended the theory.

The acceptance of Einstein's theories of relativity was also slow. He developed his special theory of 1905 with his general theory in 1916, and it is noteworthy that when he received his 1921 Nobel Prize, it was not for relativity. In 1921 the scientific world was still not entirely convinced by the theory. Today, of course, much evidence has confirmed it.

Even after scientists were convinced, there was some surprising opposition from outside of science. In the early 1930s Einstein was invited to give a lecture at Oxford on his theory. A number of philosophers were present, and to Einstein's surprise they attacked the theory on philosophical grounds. At the time relatively few people understood the theory, which is highly mathematical, and certainly the philosophers did not. Einstein explained that his theory was a scientific theory, to be judged solely by how it fit the experimental information, but most of them were unconvinced.[5]

Philosophers, sociologists, and others who attack scientific theories on nonscientific grounds fail to understand that a scientific theory is no more than an explanation of scientific information.

Chapter 3
The Ingredients of
Our Universe

In Nature's infinite book of secrecy
A little I can read.

—William Shakespeare, *Antony and Cleopatra*, Act 1 Scene 2

I f someone asks what a particular cathedral is built of, we might answer laconically "stone" and leave it at that. We know that in reality much more has gone into the cathedral than stones. Even the oldest buildings had plans of some sort. Also, much energy went into the building, and there were many intangibles such as loving care and religious faith. The same is true of our universe: it is not enough to say that it was built of certain materials. We must also think of the energy contained in the universe and the other intangibles.

Scientific investigations have led us to conclude that the universe is established according to an elegant pattern that has three distinct features. In the first place, there are certain fundamental particles out of which all matter is built (see table 1). For simplicity we will say that these are the protons, neutrons, and electrons, which it is convenient to regard as the building stones of atoms and molecules and therefore of all matter. It is true that there are many smaller particles, such as quarks, which are involved in forming the atomic nuclei, but to keep things simple we will forget about these for now.

In the second place, our universe contains energy, corresponding to the work done in constructing a cathedral. All scientific investigations have shown that energy cannot be created or destroyed, but can be converted into other types of energy, and we now know from the theory of relativity that mass is one of the forms of energy. As we shall see, this energy exists in several forms, one of which is light. Light is conveniently regarded as composed of particles called *photons*, which have the odd property of apparently having no mass when they are not moving—but they are always moving.

We must also take account of Einstein's famous equation, according to which energy and mass are interrelated by:

$$E = mc^2$$

where E is the energy, m is the mass, and c is the speed of light. This means that we should not really be discussing matter and energy as if they were quite different. In many situations they are distinguishable, but it is convenient at first to consider them separately, and later to discuss how they are related.

Third, there are certain basic and universal laws or rules of nature, such as Newton's law of gravitation and his laws of motion. There are some other fundamental principles that are always followed. One of them is that energy is quantized, that is, it comes in packets of specified sizes. Another rule of nature is the *principle of uncertainty*, which says there is a limit to how precisely we can make simultaneous determinations of more than one quantity, and which therefore places an inescapable limit on our understanding of nature. These are just a few of the universal principles or laws that we introduced in the last chapter and will discuss further in the present one. There are also some empirical laws, such as Boyle's law, which are less basic because they are the results of the universal principles.

Table 1: Particles of Nature

Particle	Relative Mass (proton mass = 1)	Relative Charge (proton charge = 1)
Electron	1/1836	-1
Proton	1	+1
Neutron	1	0

Some of the more important basic ingredients of the universe are summarized in table 2. These are the innocent-looking components of our enormously complex universe with all its wonderful galaxies, planets, oceans, rivers, lakes, mountains, and valleys, to say nothing of about 30 million different living species. Some of these species, particularly our own, are so complex that we still do not fully understand how they operate.

Table 2: Some Ingredients of the Universe

Particles	Energy	Universal Principles
Electron	Potential (e.g., gravitational)	Newton's laws of motion
Proton	Kinetic	Gravitational theory
Neutron	Heat	Three-dimensional space: inverse square law
	Work	Mass-energy relationship ($E = mc^2$) ($E = mc^2$)
	Light (photons)	Conservation of mass-energy
	Mass	Laws of thermodynamics
	(Electrical)	Boltzmann distribution
	(Nuclear)	Quantization
	(Chemical)	Uncertainty principle
		Relativity theory
		Deterministic chaos

Strictly speaking, chemical, electrical, and nuclear energy are not separate kinds of energy, but are included under the other categories. Nevertheless, it is convenient in many applications to refer to chemical and electrical energy.

The purpose of this chapter is to give some idea of the way the universe seems to have been put together from these components. That is rather an ambitious undertaking for one chapter, but I think that by looking at the situation in broad outline and making many simplifications we gain a clearer understanding of what science is really concerned with. Some of the popular misconceptions about science spring from seeing only details of narrow branches of science, and failing to see how they all fit together.

When we look at a piece of metal, it is natural at first to think of it as solid throughout. If we were to go on magnifying it indefinitely, why would it not be just the same? For centuries people assumed that to be the case, but that intuitive idea is wrong. A few of the ancient philosophers, however, notably the Latin poet Titus Lucretius Carus (c. 99–55 BCE), put forward the theory that matter was made up of atoms. Much later, Isaac Newton assumed the existence of atoms, partly to explain how light could pass through certain materials such as glass; how could it do so unless there were spaces between the constituents of matter?

It was not until the nineteenth century that scientists began to take atoms more seriously. Early in that century the English chemist John Dalton (1766–1844) showed that the atomic theory explained the fact that in chemical compounds there are certain simple numerical relationships between the amounts of the elements of which the substances were composed. For example, water always contains a certain fixed ratio of hydrogen and oxygen, which Dalton at first explained by saying that one atom of hydrogen combines with one atom of oxygen to give a molecule of water. Later it was realized that the molecule of water must contain two atoms of hydrogen and one of oxygen, and we thus write the molecule as H_2O. Later work in that century led to the conclusion that atoms and molecules are much smaller than had previously been supposed. For example, eighteen grams of water (which chemists call a mole), contains roughly 6×10^{23} molecules. It is difficult for us to comprehend such a large number, or the tiny molecular mass that this number implies. One way to get some feeling for the number is to note that every time we breathe in we take in over a million of the very same molecules once breathed in and out by any famous person of old you want to mention—Aristotle or Julius Caesar. I find this rather a breathtaking thought. Every time we drink a glass of water we can be confident that it includes many of the very same water molecules that were in the draught of hemlock with which Socrates ended his life.

We can get a still better idea of such an incredible number as 6×10^{23} in another way. Suppose we put that number of dollars in a bank that gave no interest (lucky bank). If we spent the money at the rate of a billion dollars a *second* (don't ask me how; I really have no idea, but it should be enjoyable), how long before we ran out of money? About nineteen million years. If the money were invested at only 1 percent per year, and we spent at the rate of one billion dollars a second, the capital would accumulate, not diminish.

The following may also help us to gain some feeling for how small the atoms are. Suppose we look at a ball bearing that is one millimeter in diameter; this is about one twenty-fifth of an inch, or roughly half the thickness of an average coin. Imagine magnifying the bearing until each of the atoms in it becomes one millimeter in diameter. The linear magnification required would be about ten million, and the ball bearing itself would then have a diameter of ten kilometers (roughly six miles). A glass of water with the same magnification would be about one hundred kilometers (over 600 miles) high, very much higher than any building or mountain on Earth.

There is another way in which we can appreciate the extreme smallness of atoms. A liter of water (about two pints) contains roughly 3×10^{25} molecules. Imagine stringing that number of water molecules end to end so that we have a filament of water molecules. How long would it be? The answer is about ten million million kilometers (10^{13} km or roughly 6 million million miles), which is a little more than a light-year. The filament would stretch to the Moon and back over twelve million times.[1] This estimate recalls a demonstration that the great American scientist and statesman Benjamin Franklin (1706–1790) made several times on his visits to England. In those days a man usually carried a walking stick, and it often had a top that could be unscrewed to reveal a cavity. In the head of his bamboo cane Franklin often carried what he called a "cruet of oil" and on occasion would pour a little of it on a lake. He did that, for example, on Derwent Water in the Lake District, and found that a teaspoonful of water would spread over half an acre of the lake. From those facts Franklin could easily have made an estimate of the size of the molecules; at the time, however, no one thought much about atoms and molecules.[2]

A related calculation is also surprising. Suppose that instead of stringing 3×10^{25} molecules of water together we string together that number of baseballs. That string would be about 10^{22} kilometers or 10 billion light-years, which is just about the distance to the most distant galaxies that can be observed.

Until the latter part of the nineteenth century, scientists thought of atoms as little hard balls. They seemed to carry hooks of some kind, because they had to hang on to other atoms to form molecules. Thus a water molecule, H_2O, was thought of as an oxygen atom having two hooks, each one of them attached to a hook on a hydrogen atom, which carries only one hook. To preserve a little dignity, chemists did not actually call them hooks; they called them valences (U.S.) or valencies (U.K.), but the idea was the same. Oxygen had a valence of two, hydrogen a valence of one. A pair of hooks holding two atoms together was called a valence bond, or just a bond.

This was a useful procedure, but it did not explain much. How is a valence bond really formed? We will answer this question later in the chapter, when we know more about the structure of atoms.

During the nineteenth century much work was done on the study of electricity, which led to a deeper understanding of the nature of atoms. Michael Faraday (1791–1865; see figure 5) made many advances in the study of electricity, and investigated the process of electrolysis, in which an electric current that is passed through a solution decomposes a chemical compound into its component parts. When electricity passes through water (preferably with something dissolved in it to make it conduct better), the water is split into hydrogen and oxygen, which appear as gases at the two poles. Faraday suggested that these poles, which are put into the liquid so the current can pass, should be called electrodes. Faraday's experiments soon led to the suggestion that there must be some electric particle that passes along a wire (as it later turned out, with the speed of light) when a current is passed through it. This electric particle must be able to become attached to an atom, and an atom must sometimes give up a particle. The details are a little complicated, but the main point is that there was evidence of the existence of a particle of electricity that passes along wires and somehow interacts with molecules. In 1874 the Irish physicist George Johnstone Stoney (1826–1911) discussed the problem, and he suggested that there existed a negative particle of electricity he called an electron, a name that has stuck. He even estimated the value of the charge on the electron, but had no way of estimating its mass.

Fig. 5. Michael Faraday made discoveries of fundamental importance, particularly in the fields of chemistry, electricity, and magnetism. His discovery of electromagnetic induction eventually led to the widespread distribution of electric power.

As often happens in science, further insight into the nature of electrons was obtained from experiments of a completely different kind. They involved discharge tubes—familiar to us as fluorescent lamps, such as the neon tubes used for advertising. A discharge tube is essentially a sealed glass tube with the air pumped out of it. The air is partially replaced by allowing a small quantity of, for example, neon into the tube and then electricity is passed through it. The bright discharge is partly caused by the flow of negatively charged electrons. In the final decade of the nineteenth century, the British physicist John Joseph Thomson (1856–1940) began to study the electrons in detail. With carefully designed experiments he was able to measure both the mass and charge of the electrons, and in 1897 announced that an electron was about two thousand times lighter than a hydrogen atom. He went on to investigate positively charged particles, which he found to be much heavier, their masses more or less the same as the masses of atoms.

It soon became clear that an atom is nothing like a tiny hard ball. Ernest Rutherford (1871–1937; see figure 6) deduced that an atom has a nucleus that is very much smaller than the atom itself. Surrounding the nucleus are electrons, which give the atom its size. The simplest atom is the hydrogen atom (see figure 7a), with one electron bearing a charge that exactly balances the charge of the nucleus; the atom as a whole has no charge. This particular nucleus, which bears a single positive charge, is called a proton. Although recent work has revealed that the proton is made up of smaller particles called *quarks*, it is convenient for our purposes to regard it as one of the fundamental particles. Most of the hydrogen atoms on earth have nuclei consisting of single protons, but two other forms of hydrogen exist. One of them, shown in figure 7b, has a nucleus that is about twice as heavy as for ordinary hydrogen. Its nucleus consists of a proton and a neutron, a particle that has almost exactly the same mass as the proton, but no electrical charge. We call this form hydrogen-2 or *deuterium* and give it the symbol D; we can also write it as ^2_1H, meaning that there are two particles in the nucleus, and one of them is a proton, the other being a neutron. *Heavy water*, used in some nuclear reactors, is an oxygen atom combined with two deuterium atoms, D_2O.

A proton can also become attached to two neutrons, so that its mass is three times as great as that of ^1_1H. We call this form (see figure 7c) hydrogen-3 or *tritium*, and give it the symbol T or ^3_1H. This form of hydrogen is very unstable, and does not survive very long. Atoms whose nuclei have the same number of protons but that differ in the number of neutrons are called *isotopes*. Different isotopes of an element are always very similar in chemical and physical properties, which depend much more on the charge on the nucleus than on the mass.

The nuclei of the atoms are tiny compared with the atoms themselves. Evidence for this came from experiments carried out in 1909 in the labora-

Fig. 6. Ernest Rutherford (1871–1937), 1st Baron Rutherford of Nelson, was the chief founder of nuclear physics. He made many pioneering investigations on atomic nuclei and radioactivity. This photograph was taken at McGill University, where Rutherford did the work that led to his Nobel Prize. (Courtesy of Professor John Campbell and of members of the Rutherford family.)

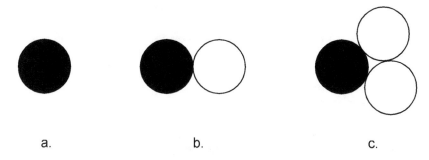

a.   b.   c.

Fig. 7. The nuclei of the three isotopes of hydrogen, showing the numbers of protons (filled circles) and neutrons (open circles): (a) Ordinary hydrogen, or protium; the nucleus is a proton, and there is an orbital electron; (b) deuterium, where the nucleus is a proton and a neutron; (c) tritium, where the nucleus is a proton and two neutrons.

In the neutral atoms, each of these nuclei is associated with a single electron. In the old theory this electron is regarded as being in an orbit, but according to quantum mechanics it exists as a kind of cloud called an *orbital.*

tory of Ernest Rutherford at Manchester University. Two of his research students, Hans Geiger (1882–1945) and E. Marsden (1889–1970), fired alpha particles (α-particles), which are the nuclei of helium atoms, at a thin sheet of metal foil and determined the direction in which they were scattered. Most of the particles were scattered by only a few degrees, but a few of them bounced right back from the foil. In Rutherford's words: "It was as though you had fired a fifteen-inch shell at a piece of tissue paper and it came back and hit you." Rutherford deduced that there was powerful electric repulsion between the positively charged α-particle, and the positively charged nucleus of the atom, and he was able to estimate the size of the nucleus. Typically, a nucleus occupies only about one million millionth (10^{-12}) of the volume of the atom. To gain some idea of the tiny size of the nucleus of an atom compared with the size of the atom itself, imagine an atom magnified to a radius of ten meters (about thirty-three feet), so that its volume is roughly that of a bus. The radius of the nucleus would be less than a millimeter—about the size of the dot at the end of this sentence. If a nucleus were magnified so that its radius was about the width of this page, the electrons would be more than a kilometer (about six-tenths of a mile) away.

The neutrons are not figments of scientists' imaginations, although that was literally the case until they were discovered and properly identified in 1932 by the British physicist James Chadwick (1891–1974). Before that, it had been assumed that atomic nuclei were composed of protons and electrons. Chadwick thought it unlikely that a proton and an electron could remain close together; he thought that instead they would combine together to form a particle, which he called a neutron. For some twelve years Chadwick looked for experimental evidence for the existence of the neutron, and in 1932 realized that there was some work done in other laboratories that provided that evidence. He quickly carried out further experiments, proving their existence beyond doubt. Although neutrons can now be detected reasonably easily, earlier detection techniques were inadequate. Since neutrons bear no electric charge, they can pass right through many types of matter without creating much of a disturbance—just as an invisible man might easily escape notice in a crowd. Neutrons can make themselves felt if they are moved at high speed with a powerful accelerator such as a cyclotron. With modern techniques, physicists can generate and experiment with neutrons, and we know a good deal about them. Certain highly condensed stars that consist almost entirely of neutrons. An interesting property of neutrons is that they are very unstable by themselves, but when combined with protons in an atomic nucleus, they are very stable. We should not think of a deuterium nucleus as containing a proton and a neutron sitting side by side, as they appear to be doing in figure 7b; they combine in a particular way that scientists are investigating and are now beginning to understand.

All nuclei can be regarded as made up of protons and neutrons, but there cannot be every possible combination of them. Certain combinations, as we will see later, cause the nucleus to be unstable. There is no known nucleus composed of just two protons, but there is one that contains two protons and one neutron (see figure 8a), and another that contains two protons and two neutrons (see figure 8b). Since there are two protons, these nuclei have a charge of plus two, and in the neutral atom each is associated with two electrons outside the nucleus. Since the chemical behavior of an atom is determined by the number of electrons required to balance the charge of the nucleus, the atoms associated with the nuclei shown in figure 8 are no longer hydrogen but another chemical element, helium. These two forms of helium, which we write as $^3_2$He and $^4_2$He, are isotopes of helium (the subscript tells us the number of protons, the superscript the total number of particles in the nucleus, also called the mass number). The proton and the neutron weigh much the same, so $^4_2$He is about four times as heavy as an ordinary hydrogen atom, while $^3_2$He is three times as heavy. About 99.999 percent of the helium found in nature is $^4_2$He, so the amount of $^3_2$He is exceedingly small.

Carbon is particularly important in living systems, and it exists in three known isotopic forms. The most common, with an abundance of 98.9 percent, is $^{12}_6$C or carbon-12, whose nucleus consists of six protons and six neutrons, so that the mass number is $6 + 6 = 12$. There also exists $^{13}_6$C, carbon-13, in which there is an extra neutron; its abundance on the earth's surface is 1.1 percent. A third form, $^{14}_6$C, has also been detected in extremely small amounts.

For the lighter elements, a useful rule of thumb is that the most common isotopes have equal numbers of protons and neutrons. We see

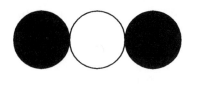

a.	b.

Fig. 8. Representation of the nuclei of the two known isotopes of helium, showing the numbers of protons (filled circles) and neutrons (open circles). Two neutrons (b) is by far more common. A fact that is significant to the origin of elements is that isotopes $^{2}_{2}$He and $^{5}_{2}$He do not exist. The former would have two protons and no neutrons in its nucleus, which would be highly unstable. The latter would have two protons and three neutrons, which again is too unstable an arrangement.

that this is true for oxygen, where the most common isotope is $^{16}_{8}$O, with eight protons and eight neutrons. It is also true for nitrogen, where the commonest form is $^{14}_{7}$N, with seven protons and seven neutrons. When we get to the heavier elements, however, this is no longer the case. The only isotopes that are at all stable tend to have many more neutrons than protons. Here are some examples:

Radium $^{226}_{88}$Ra   88 protons and 138 neutrons; 226 particles

Thorium $^{232}_{90}$Th   90 protons and 142 neutrons; 232 particles

Uranium $^{238}_{92}$U   92 protons and 146 neutrons; 238 particles

Although these isotopes are the most stable of their respective elements, they do not exist forever, unlike many of the lighter elements such as carbon and oxygen, which are completely stable in at least one of their isotopic forms. The three elements, radium, thorium, and uranium, all show radioactivity, a phenomenon that was discovered in 1896 by the French physicist Antoine Henri Becquerel (1825–1908). He found that uranium, the most abundant radioactive element in nature, emits α-particles, the nuclei of $^{4}_{2}$He atoms; the process is called *radioactive disintegration*. We can understand it with a little arithmetic:

$^{238}_{92}$U (92 protons + 146 neutrons) $\rightarrow$ $^{4}_{2}$He (2 protons + 2 neutrons) + what?

The end product must contain 92 – 2 = 90 protons, which means that it is the element thorium (the name of the element is determined by the number of protons in the nucleus). It has 144 neutrons, the total number of particles being 234:

$$^{238}_{92}U \text{ (92 protons + 146 neutrons)} \rightarrow {}^{4}_{2}He \text{ (2 protons + 2 neutrons)}$$
$$+ {}^{234}_{90}Th \text{ (90 protons + 144 neutrons)}$$

We see that the thorium produced in this process is not the most stable form, which was shown earlier, $^{232}_{90}Th$; it has two neutrons too many. It also undergoes radioactive disintegration, but not in the same way as uranium. It does not give off an alpha-particle, which is a helium nucleus, but instead emits a beta particle (β-particle), which is an electron. Emission of an electron is equivalent to converting a neutron into a proton. A neutron can be written as $^{1}_{0}H$, since it has no protons and a mass of 1. A proton, with one proton and a mass of one, is $^{1}_{1}H$. The process is thus:

$$^{1}_{0}H \rightarrow {}^{1}_{1}H + \text{what?}$$
$$\text{(neutron) (proton)}$$

In order to make the numbers in this equation balance, the subscript must be –1 meaning that the charge is –1, while the superscript is 0, which means that the tiny mass is effectively zero. Since the particle emitted was originally called a beta particle, we use the symbol β and write the electron as $^{0}_{-1}β$. The equation for the conversion of a neutron into a proton is thus

$$^{1}_{0}H \rightarrow {}^{1}_{1}H + {}^{0}_{-1}β$$
$$\text{(neutron) (proton) (electron)}$$

When a $^{234}_{90}Th$ (90 protons + 144 neutrons) nucleus emits a β-particle, the product nucleus has 91 protons and 143 neutrons, the total number remaining at 234, and we can write the process in terms of the balanced equation:

$$^{234}_{90}Th \text{ (90 protons + 144 neutrons)} \rightarrow {}^{0}_{-1}β \text{ (electron)} + {}^{234}_{91}Pa$$
$$\text{(91 protons + 143 neutrons)}$$

The element that has a nuclear charge of 91 is called protoactinium, Pa.

These radioactive processes are typical of many other spontaneous nuclear processes. In some of them a helium nucleus (alpha particle) is emitted, in others an electron. Radioactive disintegrations also differ considerably in the rates with which they occur. A nucleus undergoes radioactive disintegration so that half of it disappears after a certain time (see

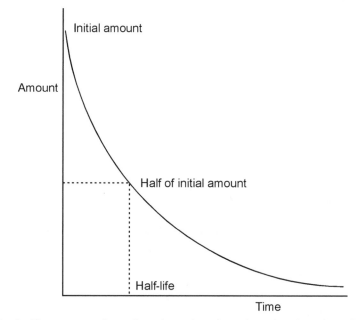

Fig. 9. The amount of a radioactive substance, plotted against time. For a given radioactive nucleus, however much we start with, half of it will have disappeared after a certain time.

figure 9). This time is known as the *half-life*, and it has a particular value for each nucleus, varying widely from one nucleus to another. An important feature of the half-life is that it is completely independent of outside factors, such as temperature.

The half-life of ordinary uranium, $^{238}_{92}U$, is about 4.5 billion years. That means that if we have ten grams of it now, we will have five grams of it after 4.5 billion years have elapsed; the rest will have decomposed into thorium and helium. The isotope of thorium produced in the disintegration of uranium, $^{234}_{90}Th$, has a much smaller half-life, only 24.1 days. The common form of radium, $^{226}_{88}Ra$, has a half-life of 1,620 years. Some isotopes have half-lives of a tiny fraction of a second, and are therefore very hard to study experimentally. An isolated neutron disintegrates into a proton and an electron (a β-particle) within its half-life of 15.3 minutes. As we have seen, however, the neutrons in atomic nuclei are much more stable.

A particularly important constituent of our universe is energy.[3] It is convenient to consider this as a separate constituent, even though we know that, according to Einstein's special theory of relativity, energy is related to mass. It may be that before the universe was created there was no matter and everything was energy. If this is true, creation started with the conversion of energy into matter, and since that time there have been many interchanges between mass and energy.

Before we go into more detail, it is best first to think of energy as a separate entity, which is justifiable as long as we are not concerned with nuclear processes. When an ordinary chemical reaction takes place, even if it is an explosion and produces large amounts of energy, the decrease in mass that occurs is too small to be measured by ordinary techniques. The same is true of industrial processes—as long as they are not nuclear. Scientists and engineers who are concerned with ordinary, nonnuclear processes can therefore forget about the energy-mass interconversion.

Although today we talk a lot about energy, it is a comparatively new word. It does not appear in the Bible, and Shakespeare never used it. The poet William Blake tells us that according to the Devil, "Energy is Eternal Delight," and Alexander Pope makes a passing reference to "energy divine." But even Isaac Newton never seems to have used the word, or if he did, he did not make much of it. Until the nineteenth century, in fact, the word *energy* was rarely used at all, even by scientists. This is somewhat surprising, because we can take Newton's equations of motion and, by simple algebra, convert them into an equation relating two kinds of energy, kinetic and potential. Today we often think in terms of energy, and it is important, because it cannot be created or destroyed, as we discussed in the last chapter (the first law of thermodynamics).

The simplest kind of energy to understand is what is called *kinetic energy*, the energy that something possesses by virtue of its motion. The expression for the kinetic energy of a body of mass m moving with a speed of v, is

$$\tfrac{1}{2}mv^2$$

A body also has energy by virtue of its position, and we call this kind of energy *potential energy*. The special case of energy resulting from the force of gravity is called *gravitational energy*. Thus, if we lift a stone of mass m from the ground to a height d, we say that its potential energy or gravitational energy has increased by

$$mgd$$

where g is the *acceleration of gravity*. Its value is much the same wherever we are on Earth, but if we go to the much lighter Moon, it is much less. The product mg is actually the weight, or force, exerted by the mass when it is subjected to the gravitational attraction of the earth. This potential energy mgd is equal to the work we have to do in raising the body to the height d.

It is easy to show from Newton's laws that if we raise a body to a height d and then release it, its speed when it reaches the ground will be obtained by equating these two expressions for the kinetic energy and the potential energy. In other words, potential energy and kinetic energy can be interconverted. This is just one example of the important principle that energy is always conserved (always aside from the interconversion of mass and energy according to relativity theory). Incidentally, this explains why we define kinetic energy as $\frac{1}{2}mv^2$ rather than simply mv^2, as was done at one time; it is so that potential energy and kinetic energy just balance each other.

Every type of energy can be classified as kinetic energy, potential energy, or work. Work is done whenever there is movement of a body against a resisting force. If we push against a heavy piece of furniture but fail to move it, we are doing no work on it, whatever we may think about all our effort. If we push harder and succeed in moving it, we are then doing work against the resisting force of friction. If, in moving the furniture, we exert a constant force and push it for a certain distance, the scientific definition of the work is that it is equal to the force multiplied by the distance.

There is no loss or gain of total energy; energy is just converted from one form into another. When we drop something and it lands on the ground and stays there, we may think that energy has disappeared; it had kinetic energy $\frac{1}{2}mv^2$ as it hit the ground, but then has none. Actually the energy has been converted into other forms, primarily heat. With delicate enough instruments we would be able to detect that the surrounding objects had become a little warmer.

We discussed heat in the last chapter, particularly in connection with Joule's experiments. He showed that work is always converted into an equivalent amount of heat and so proved experimentally the first law of thermodynamics, which is essentially the same as the principle of conservation of energy.

In dealing with chemical reactions, chemists make much use of what they call *chemical energy*, which, although not a distinct form of energy, is a convenient concept. When most (but not quite all) chemical reactions occur, heat is liberated and the system gets warmer. An extreme case of this is an explosion. The surroundings get warm, and in addition, some material damage is generally done. This movement of bodies means that work has been done. It is convenient to say that chemical energy was stored in the chemical system and is released as the explosion takes place. If we investigate these systems carefully, we find that there is never any loss or gain of energy in a chemical process; the chemical energy has been converted into other kinds of energy, a necessary consequence of the first law of thermodynamics. This is something we are all familiar with. If we heat our homes with oil, a furnace allows the oil to undergo a chemical reaction (burning, which is oxidation), and the energy is released as heat. Strictly speaking, chemical energy is not a distinct form of energy, but a corresponding change in the potential energy when a chemical reaction occurs, due to the reorganization of the nuclei and electrons.

Similarly, we often speak of *electrical energy*, although this also is not a distinct form of energy. When a current flows, there is movement of electrons along conductors such as metal wires, and if batteries or cells are involved, there is movement of positive or negative particles, which we call *ions*. The energy associated with these movements can be called electrical energy. It can be converted into heat, something we bring about if we heat something on an electric stove. Recall that Joule demonstrated that there is a one-to-one relationship between electrical energy and the heat it produces.

We can also speak of *light energy*. When light interacts with an atom, it may provide energy and give it to an electron, which moves into another state where it is farther from the nucleus and therefore more energetic. Light can sometimes actually eject an electron from an atom; this is known as the *photoelectric effect*. To explain this effect we must regard light as consisting of *photons*, which are tiny bundles of energy that can interact with atoms and transfer their energy, being annihilated in the process.

Here are the fundamentally different forms of energy we have considered so far: kinetic, potential (e.g., gravitational), heat, and work. We have also mentioned the energies associated with chemical reactions, electricity, and light. We now come to nuclear energy, which, under special circumstances, is produced in much larger amounts than any other kind of energy.

The interconversion of energy and mass does not affect our everyday lives, but it becomes important when we are concerned with nuclear transformations. Since these occur in the sun and all stars, they are important for understanding the ultimate sources of energy. In particular, our deductions about how the universe must have been created require us to understand mass-energy transformations, as expressed by Einstein's formula $E = mc^2$. This equation was a consequence of Einstein's special theory of relativity.

To get some idea of the enormous amounts of energy released when mass is converted into energy, suppose that you were skiing down a hill at a high speed. Suppose, hypothetically, that you attained the speed of light. It follows from Einstein's formula that if you hit a tree, the liberation of energy due to your complete annihilation would be equivalent to about forty nuclear warheads. There is another way of appreciating the enormous amounts of energy that are tied up in nuclei. If all the energy of one ounce of matter could be released, it would be enough to drive the largest liner across the Atlantic.

One example of how mass can be converted into energy is provided by one of the radioactive disintegrations that we considered earlier. Ordinary radium undergoes the following process, the products being the elements helium (He) and radon (Rn):

$$^{226}_{88}\text{Ra} \quad \rightarrow \quad ^{4}_{2}\text{He} \quad + \quad ^{222}_{86}\text{Rn}$$

The following are the masses of the nuclei involved:

$$^{226}\text{Ra} = 226.0312 \text{ u} \qquad ^{4}\text{He} = 4.0026 \text{ u}$$
$$^{222}\text{Rn} = 222.0233 \text{ u}$$
$$\text{sum} = 226.0259 \text{ u}$$

The unit, "u," is the *atomic mass unit*; it is defined so that the mass of the isotope $^{12}_{0}\text{C}$ (carbon-12) of the carbon atom is exactly 12 u. In the process above, the total mass decreases by 226.0312 – 226.0259 = 0.0053 u. Calculating from Einstein's formula, one atomic mass unit corresponds to an energy of 931.5 million electron volts, MeV, so a decrease in mass of 0.0053 u leads to the production of 4.9 MeV of energy. The *electron volt* is the energy acquired by an electron when it passes through a voltage drop of one volt. One electron volt is roughly the energy that is carried by one photon of visible light. We get some idea of the vast amount of energy that is emitted in these nuclear transformations when we note that in ordinary chemical reactions, even in conventional explosions, the energy produced is only a few electron volts. With nuclear processes, however, we are talking about millions of electron volts. The energy released when a conventional nuclear bomb explodes is roughly a million times greater than when a chemical bomb of similar size explodes.

We have seen that α (alpha) and β (beta) particles are often emitted in radioactive decay. Remember that an alpha particle is the nucleus of a helium atom, $^{4}_{2}\text{He}$, while the beta particle is an electron, $^{0}_{-1}β$. We often refer to a beam of these high-energy particles as radiation, and we must be careful not to confuse this *particle* radiation with *electromagnetic* radiation, of which ordinary light is an example. A third type of emission from radioactive processes is electromagnetic radiation, of much higher energy than visible light and even x-rays. It is called gamma (γ) emission. The reason it is emitted in a radioactive disintegration is that the nucleus produced in the process we just considered, radon (Rn), is usually not formed in its most stable form, but with excess energy. It rapidly becomes converted into the most stable form, with the emission of energy in the form of gamma (g) radiation.

There are two other important types of nuclear transformation. One is called *nuclear fission*, and involves the breakdown of a nucleus into smaller nuclei. The first nuclear fission reaction was brought about in the Cavendish laboratory at Cambridge by two of Rutherford's assistants, John Douglas Cockcroft (1897–1967) and Ernest Thomas Sinton Walton (1903–1995). They introduced a neutron into uranium-235. It was captured, and formed uranium-236, which breaks down into two nuclei, barium-145 and krypton-88. Since 145 + 88 = 233 is three less than 236, there are three neutrons left over. Since the neutron has unit mass but no charge, we write it as $_0^1 n$, and the process is:

$$^{235}_{92}U + {}^1_0 n \rightarrow {}^{236}_{92}U \rightarrow {}^{145}_{56}Ba + {}^{88}_{36}Kr + 3{}^1_0 n + \text{energy}$$

The amount of energy released in this process is enormous, about 200 MeV, and some of it appears as gamma radiation. What is particularly important about this reaction is that a single neutron has produced a large amount of energy without becoming lost; more than that, three neutrons have appeared in its place. These neutrons will bring about the breakdown of other uranium nuclei, with the production of even more neutrons. This is called a chain reaction, and within a tiny fraction of a second many of the uranium nuclei present will have undergone fission. A small number of neutrons introduced into uranium-235 therefore lead to the production of vast amounts of energy. In practice, less energy is released than if all of the nuclei underwent fission; only a few percent of the energy that is latent in the U-235 is released. Nevertheless, the energy produced is vastly more than in an ordinary chemical reaction, about a million times as much.

The first atomic bomb ("Little Boy") that was dropped on Hiroshima, Japan, in August 1945 was based on this process. Ordinary uranium is mainly uranium-238, which is nonfissionable. To make the uranium suitable for a nuclear bomb, it had to be enriched in uranium-235, a process that took much time and effort. Uranium-235 is a natural emitter of neutrons, but if its amount is less than a certain "critical mass," the chain reaction does not take place. To detonate the bomb, two subcritical masses of uranium enriched in uranium-235 were shot together suddenly by means of ordinary explosives, producing a critical mass that produced a stupendous explosion. The bomb was used at Hiroshima without any prior test; only small amounts of uranium rich in uranium-235 had been separated, and it was not thought desirable to waste any on a test.

During World War II, physicists became aware of another possible route to an atomic bomb. Uranium-238, which is much more abundant than uranium-235, cannot sustain a chain reaction associated with fission. However, when uranium-238 captures a neutron, it is transmuted in three successive stages into another element, which is called plutonium, Pu. With the omission of an intermediate stage, the overall process is:

$$^{238}_{92}U + {}^1_0 n \rightarrow {}^{239}_{92}U \rightarrow {}^{239}_{94}Pu + 2{}^0_{-1}\beta \text{ (beta rays)} + \text{gamma rays}$$

Note that there is emission of both beta rays, which are electrons, and gamma rays, which are electromagnetic.

If it is bombarded by neutrons, a plutonium-239 nucleus undergoes fission in much the same way as uranium-235, and again vast amounts of energy are liberated. Since plutonium is a distinctly different element, rather than an isotope of uranium, it can more easily be separated from uranium by chemical means. The nuclear reactor in World War II that used uranium-238 was thus a producer of plutonium, from which another type of atomic bomb was made.

The nuclear bomb "Fat Man," dropped over Nagasaki three days after "Little Boy," was a plutonium bomb. Its plutonium was produced in a nuclear reactor, using controlled fission of uranium-235 to produce the neutrons that caused uranium-238 to produce plutonium-239.

Another kind of nuclear process, called *fusion*, in principle leads to the production of even greater amounts of energy. In a fusion reaction the nuclei simply combine together with the liberation of vast amounts of energy. An important example involves deuterium. The mass of the nucleus of a deuteron ($_1^2H$) is 2.0141 u, while that of a helium atom ($_2^4He$) is 4.0026 u. Therefore, if two deuterons form a helium nucleus, there is a mass loss of 4.0282 − 4.0026 = 0.0256 u, and a corresponding release of energy of about 23.8 MeV:

$$_1^2H + {}_1^2H \rightarrow {}_2^4He + 23.7 \text{ MeV}$$

This reaction took place extensively in the big bang, at the creation of the universe, and subsequently has occurred in the stars.

It is difficult to make this reaction occur on Earth, because the highest temperatures that can be attained are much lower than those in the big bang or in the stars, and are too low for much $_2^4He$ to be formed. Instead, the following two reactions, which produce less energy, occur instead:

$$_1^2H + {}_1^2H \rightarrow {}_2^3He + {}_0^1n + 3.28 \text{ MeV}$$

$$_1^2H + {}_1^2H \rightarrow {}_1^3H + {}_1^1H + 4.04 \text{ MeV}$$

These reactions occur in the so-called hydrogen bomb, or *thermonuclear bomb*. Special techniques are necessary to produce extremely high temperatures, which are called thermonuclear. In the stars the temperatures are sufficiently high that the fusion reaction to form $_2^4He$ goes on all the time. The enormous amounts of energy released in fusion reactions keeps them hot for billions of years, but eventually a star will die when its nuclear fuel is used up.

It is interesting to note that the reaction that led to James Chadwick's discovery of the neutron (see pages 63–64) is a fusion reaction. The experiments involved the bombardment of beryllium by alpha particles, and Chadwick showed that the process is:

$$_2^4He \text{ (alpha particle)} + {}_4^9Be \rightarrow {}_6^{12}C + {}_0^1n \text{ (neutron)}$$

Now we should look at some other intangibles, the universal laws or general scientific principles. Our real understanding of nature's laws may be said to have started with Newton's three laws of motion, set out in his *Principia Mathematica*, published in 1687. Today they seem rather obvious, and we may be tempted to wonder why it took so long for them to be discovered; however, that is because we are quite used to them in everyday life. They are as follows:

1. A body persists in its state of motion unless it is acted upon by a force. This principle had previously been stated by Galileo. It seems obvious today that, for example, spacecraft keep on going without power once they are free from the earth's gravity, but that is only because we have been told that it is so. It cannot really be obvious, because Aristotle had a quite different idea about motion.
2. Newton's second law says that that if we apply a force to a body it will accelerate, and that its acceleration is greater the greater the force, and smaller the smaller the mass of the body. We express it by the equation:

$$\text{Force} = \text{acceleration} \times \text{mass}$$

3. Every action has a reaction, which is equal and opposite to it. Again, this is easy to understand. If we push on a wall, the wall pushes back on us with just the same force.

Before he formulated his three laws of motion, Newton stated his law of gravitation. One matter that troubled him for many years in formulating his theory was whether it was correct to regard the mass of a body (such as a sphere) as if it all resided at the center of the sphere; in other words, in making the calculations, is it acceptable to pretend that all the mass is concentrated in a point at the center of the sphere? It is acceptable; it is not easy to prove that it is, but Newton succeeded. Newton's law of gravitation stated that the force of attraction between two spheres is the product of their masses, and is inversely proportional to the square of the distance between the centers of the spheres. This means that if there is a certain attraction when two bodies are one meter apart, if we double the distance, the force goes down by a factor of two squared, which is four. This *inverse square law* applies to many things besides gravity. It applies to the force between two charged bodies. It applies in a rough way to everyday situations, like how clearly we can see lighted objects, and how well we can hear sound (aside from complications due to reflection).

For practical purposes we can say that we live in three-dimensional space. The inverse square law is compatible with three dimensions and would be incompatible with space of any other dimensions. We can understand this by considering a beam of light spreading out from a point (see figure 10). If a screen is at one position, the image is of a certain size, but if the screen is twice as far away from the source, the image is four times as large. Since the area of the larger image is four times the area of the smaller one, the intensity of the light is one-quarter. This is the inverse square law, and it follows from the three-dimensional nature of our space. So we can include three-dimensional space as one of the ingredients of nature.

In the early years of the twentieth century, Albert Einstein (1879–1955; see figure 11) proposed his special and general theories of relativity, which for present purposes we will consider as a modified formulation of the laws of mechanics. People often think that Einstein's laws of nature supplanted Newton's. This, however, is true only in a formal sense. From the practical point of view, Newton's laws are almost always adequate. An engineer

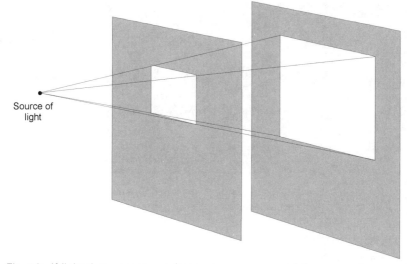

Source of light

Fig. 10. If light throws a square image on a screen, and the screen is moved to twice the distance, the image becomes four times as big. The intensity of the light at the second position is thus one quarter of that at the first position. This shows that the inverse square law is a necessary consequence of the fact that space is three-dimensional.

designing a modern car, for example, uses Newton's mechanics and ignores Einstein's. Even the design of a spacecraft is done without recourse to Einstein's laws.

The point is that we need to take Einstein's theory into account only in special and unusual circumstances. For example, if we are concerned with speeds approaching the speed of light, which is about 300,000,000 meters per second (300,000 kilometers per second or 671,100,000 miles per

Fig. 11. Albert Einstein is best known for his theory of relativity, but his work on the quantum theory was probably of equal importance. (Portrait by Karsh of Ottawa, courtesy of the Edgar Fahs Smith Collection, Annenberg Rare Book and Manuscript Library, University of Pennsylvania.)

hour), relativity theory is very relevant. After the theory was formulated, it could only be tested after a number of years, when certain astronomical events had occurred.

Most scientists go through their entire research careers without making any use of Einstein's theory. In my own research in physics and chemistry I never had to use it, and I only know a few situations where relativity has to be introduced into a purely chemical problem. One of them relates to the color of gold. When theoretical calculations were first made on the structure of solid gold, the results suggested that gold should have the same color as silver. Obviously something was wrong with the calculations, and it was traced to the fact that in gold, because of the high mass of the atomic nucleus, the electrons move at speeds approaching that of light. A relativity correction was therefore necessary, and when it was introduced, it was possible to understand why gold looks different from silver. That is one of the few things that I know that relates relativity theory to our everyday lives.

*　　*　　*

Another important general principle that applies to our universe is expressed in the quantum theory. This concept, introduced in 1900 by the German physicist Max Planck (1858–1947; see figure 12), and Einstein's theory of relativity are two of the most important scientific contributions of the twentieth century. Like relativity, the quantum theory has been confirmed so many times that we can be quite sure that it is a universal law.

Just as relativity theory does not often enter our daily lives, the quantum theory is hardly ever obvious. According to the quantum theory, energy has to come in packets, and we cannot have any amount we demand—just as in a store we are not allowed to buy any amount of milk, but have to buy it in pints or liters, and we are not allowed to pay for it in fractions of the smallest coin. What is different about the energy situation is that the packets are so incredibly small that it is hard to know that they exist. For example, when we drive a car, it is impossible for us to know that the speed at which we drive is quantized, and that only certain speeds are allowed to us. The permitted speeds are so fantastically close together that the most careful mechanical measurements would never detect that there is any quantization. In view of this, it is amusing that journalists, politicians, and others are fond of talking about "quantum leaps," not realizing that in everyday life, these "leaps" are quite undetectable. The next time you hear a politician promising a quantum leap, remember that changes from one quantum state to another are miniscule and occur with completely unpredictable results.

We can sum up the situation by saying that the so-called classical theories—Newton's laws of motion, gravitation, and so forth—apply when we are dealing with things on a large scale, which includes what we experience in everyday life. Quantum laws take control only at the atomic and molecular levels. The way time is involved is also interesting. When gravity and the classical laws take control, everything tends to proceed slowly, whereas a quantum transition occurs in a flash. For instance, in the creation of the universe, many of the quantum effects involving primordial particles were over in a tiny fraction of a second, while the creation of galaxies by gravity took billions of years,

When Planck introduced his quantum theory in 1900, he thought it applied only to the energy of vibration of atoms. In 1905 Einstein published his first paper on relativity theory, and in the same year he introduced another highly original idea that is just as important as relativity theory. He suggested that light itself is quantized, that is, that it comes in

Fig. 12. Max Planck is the chief originator of the quantum theory. In 1900 he explained the distribution of energy in radiation by the idea that energy comes in small packets, or quanta. This was so revolutionary that Planck had difficulty in believing it himself at first, but was later convinced by Albert Einstein, who showed that the theory applies to many other things besides radiation. (Courtesy of the Edgar Fahs Smith Collection, Annenberg Rare Book and Manuscript Library, University of Pennsylvania.)

tiny packets—later called *photons*. The theory of light has had a complicated and troubled history. There have been two theories about the nature of light. One was that light is a stream of tiny particles, which used to be called corpuscles. The other was that it has wave motion, involving vibrations that are at right angles to the direction of propagation of the light (see figure 13). Corresponding to the wave there is a particular wavelength, denoted by the Greek letter lambda (λ), and a frequency, denoted by nu (ν). The two are related to the speed of light by the equation:

$$\text{Speed of light } (c) = \text{wavelength } (\lambda) \times \text{frequency } (\nu)$$

Newton gave careful consideration to both the corpuscular and wave theories, and did not completely discard either of them. Indeed, he sometimes favored ideas that were partly wave and partly corpuscular, which is rather

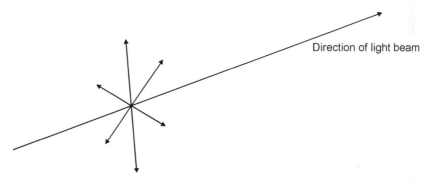

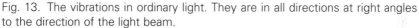

Direction of light beam

Fig. 13. The vibrations in ordinary light. They are in all directions at right angles to the direction of the light beam.

remarkable, since that is the modern view. Newton's main difficulty with the wave theory was that he could not see how it explained the fact that light travels in straight lines; he thought that when light encountered an obstruction, the waves would be sent out in different directions. There is indeed some spreading of this kind (called *diffraction*), but the spreading is very slight and hard to detect. Newton changed his views from time to time, but on the whole he preferred the corpuscular theory. Because of his great prestige, this was the theory generally accepted until the beginning of the nineteenth century.

Early in that century the British investigator Thomas Young (1773–1827) did important work on color perception by the human eye and also on the theory of light. He discovered that when light passes through narrow slits, bands of light, known as *diffraction bands*, were formed. These results can only be explained by the wave theory.

What Einstein did in 1905 was to establish that both theories are correct, and that we must have a dual theory of light. Whether light shows particle or wave properties depends on the type of experiment that is being carried out. Einstein deduced that if the frequency of light is ν (nu), the amount of energy in each particle (a photon, or a light quantum) is equal to a constant h (known as the *Planck constant*) multiplied by the frequency ν:

$$\text{Energy of photon} = h\nu$$

It is for this work that Einstein belatedly won his 1921 Nobel Prize, and not for his theory of relativity. Of course he should have gotten a second Nobel Prize for relativity.

At first, many scientists (including Planck himself) had misgivings about Einstein's theory. In the first place, the results on the diffraction of light seemed to establish the wave theory without question; how could a particle theory explain it? Also, there seemed to be a logical difficulty with Einstein's theory, according to which the energy of the photon is $h\nu$. The idea of frequency (ν) makes sense only in terms of the wave theory. Einstein agreed that the wave theory was well established and could not be cast aside. The behavior of light in a diffraction experiment relates to its *average* properties over time, and wave properties are important. However, he argued, when light interacts with an atom, there is a one-to-one effect, and what will then be important are the particles of light rather than the waves. This was something of a new idea, that different theories of light were required to explain different kinds of optical experiments, but scientists soon got used to it.

The way particle theory explains the interaction of light with an atom is shown in figure 14. An electron moving in one orbit can move into another orbit that is further from the atomic nucleus. To do this it must gain energy by absorbing a photon of light, of such a frequency ν that $h\nu$ is equal to the energy required. This idea is essential to the understanding of why the spectra of atoms show sharp lines—a question that was very puzzling before the quantum theory was introduced. If we throw some salt into a fire, the flames become bright yellow. It we look at the spectrum of a sodium flame with a simple spectroscope, we observe a sharp line in the yellow region. Why is the light emitted only at this frequency and not others? It is because only certain orbits are allowed by the quantum theory. The sodium atom happens to have two electron orbits differing in energy by an amount that corresponds to this yellow line.

Although the quantum theory has wide implications for anyone who works on the behavior of individual atoms and molecules, it was introduced by Planck in 1900 solely to explain the experimental results on the emission of radiation by hot bodies. This is of great importance to the for-

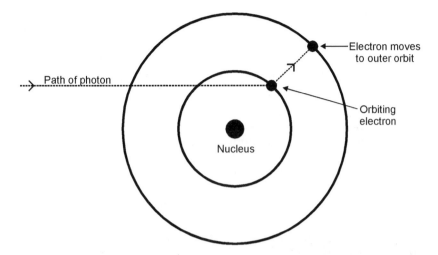

Fig. 14. The interaction of a photon of light with an orbiting electron, raising it to an orbital in which it has more energy. The quantum condition restricts the electron to certain orbits, two of which are shown. If an electron is in an outer orbit and drops to an inner one, there is emission of radiation. The frequency ν of the absorbed or emitted radiation is determined by the fact that the energy $h\nu$ must be equal to the difference between the energies corresponding to the two levels.

mation of the universe, especially since it introduces us to yet another of the universal laws about the way energy is distributed.

Planck's special interest in 1900 was in how the intensity of radiation emitted by a hot body varies with the wavelength and temperature. When an incandescent electric light bulb is controlled by a dimmer switch and as the light is gradually turned up, the tungsten filament at first emits a dull red glow. If the current to the lamp is steadily increased, the radiation becomes bright red, then orange, then yellow. At the highest temperature the light appears to be white, which means that light is emitted over a range of wavelengths. The higher the temperature, the more light is emitted at higher frequencies.

We can imagine a substance that would absorb all of the radiation that falls on it. Since it reflects no radiation, it appears black, and such a substance is known as a *blackbody*. No such material actually exists, but a close approximation is a cavity in a metal with a small hole leading to it. Any radiation passing into the cavity would have little chance of being reflected out.

During the latter years of the nineteenth century, scientists made many experimental studies of the energy of the radiation emitted by a hot blackbody. They were particularly concerned with the way the energy varied with the frequency of the radiation. A plot of the energy against the frequency always took the shape shown in figure 15, the curve passing through a maximum. What is remarkable is that, whatever the temperature, the shape of the curve is exactly the same. We show only one curve in figure 15—the amount of radiation plotted against the frequency divided by the absolute temperature. Consider the curve relating to the temperature of 5,000 kelvin (roughly 8,500 °F), which we write as 5,000 K (not as 5,000 K, the convention now being to omit the degree sign for degrees kelvin). The visible part of the spectrum is in the region of the maximum of the distribution curve. This tells us that if a body is at that temperature, a lot of radiation is given off in the visible range, and the body appears white hot. This is obviously true of the sun, the surface temperature of which is somewhat over 6,000 K (about 10,300 °F). At 2,000 K (about 3,140 °F), on the other hand, the visible part of the spectrum corresponds to only the tail end of the curve, so that little visible radiation will be given off; our eyes will detect little light. Most of the radiation will be in the infrared, and if we put our hand near the body, we will feel the heat. What little light we see will be red, because most of the visible radiation will correspond to the red end of the visible spectrum.

If we go right down to 3 K, all of the radiation given off will be in the microwave region of the spectrum, that is, in the region where the photon energies are very small. With the naked eye we will see nothing. This temperature of 3 K is of very special interest in connection with the birth of the universe, because it is the temperature of most of outer space. As we shall see, radiation corresponding to this temperature can still be observed at microwave frequencies; it is "fossil" radiation, originally emitted in the early stages of the life of the universe. The distribution of frequency corresponds to the present temperature of outer space, which is about 3 K.

The theory that, together with quantum theory, explained this behavior was proposed in 1868 by the Austrian physicist Ludwig Boltzmann (1844–1906). Boltzmann's theory is another important law of nature. It was a generalization of a treatment given earlier by the Scottish physicist James Clerk Maxwell (1831–1879). Maxwell considered the speeds that molecules move about in a gas. Some move slowly and others more rapidly, and Maxwell worked out a mathematical treatment of how the number moving at a particular speed varies with the speed. His result corresponds to a curve of a similar shape to that shown in figure 15. At the maximum of the curve, a certain number of molecules are moving at those intermediate speeds. Fewer are moving at higher and at lower speeds.

Boltzmann translated Maxwell's speeds into energies, then extended the idea to any other kind of energy. He proved that the chance that any system will have a particular amount of energy E is proportional to a fraction:

$$\exp\left(-E/k_B T\right)$$

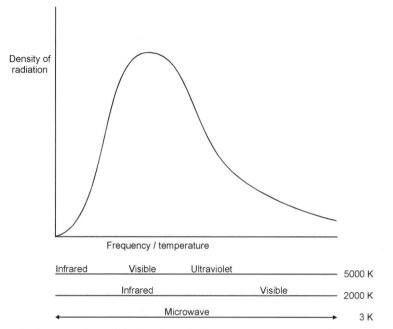

Fig. 15. The density of blackbody radiation at various frequencies. Density is plotted against the ratio of the frequency and the absolute temperature, and the curve is exactly the same whatever the temperature. The three scales at the foot of the diagram show the regions of the spectrum that correspond to three temperatures.

This requires a little explanation. When we write "exp," we mean that we are raising a mysterious little number e, which has the value 2.718281 . . . to the power of what is in parentheses. T is the absolute temperature (i.e., it is in degrees Kelvin), and k_B is the *Boltzmann constant*. This constant is of tremendous importance, as it applies to every kind of energy when there is equilibrium. The constant k_B, like the Planck constant h, is another universal constant. Planck explained blackbody radiation by modifying Boltzmann's theory, introducing the requirement that the energies are quantized.

Another universal law is that light is electromagnetic—that is, it has electric and magnetic components. The elucidation of this characteristic of light proved a matter of some difficulty, and became a matter of great practical importance. The many technical advances made in radio transmission, with all its consequences, such as television and radar, could not have come about without the electromagnetic theory of radiation worked out by Clerk Maxwell.

To gain some idea of the theory, we should go back to a famous experiment carried out in 1820 by Hans Christian Oersted (1777–1851), professor of physics at the University of Copenhagen. This experiment has nothing to do with light, but it later had important consequences for the theory of light. Oersted brought a compass needle near to a wire along which an electrical current was flowing. He found that the needle was deflected in an unexpected way; the poles of the magnet were not just attracted or repelled by the current, but moved in a direction at right angles to the expected direction. When the direction of the current was reversed, the needle turned in the opposite direction. Incidentally, there were two unusual features of this famous discovery. One is that it was made in front of a class of students; Oersted had never previously tried the experiment. The other is that, since the deflection of the needle had been small, Oersted was not impressed by the result and performed no further experiments on the subject for three months. Soon after he published his result he became famous. Yet another remarkable feature of the work was that, to ensure wider recognition in many countries, Oersted's first announcement of it was in Latin. It was soon translated into a number of languages.

This experiment showed that there is some previously unrecognized relationship between electricity and magnetism. Various theories were proposed, but Michael Faraday (1791–1867) arrived at the most fruitful interpretation. Rather ironically, one of the reasons for his success was that, because of his limited education, he was not enough of a mathematician to understand alternate theories. In their place, Faraday put forward a fairly simple explanation that many scientists, especially the good mathematicians, ridiculed because of its apparent naïveté. But Faraday was right, and their ideas were unproductive.

Faraday's idea was that, around a wire carrying a current, there are circular lines of force that constrain a magnet placed near it to behave as it does. This was later expressed by saying that there is an electromagnetic field around the wire, and modern theories in advanced physics, including Einstein's theory of relativity, make much of these field theories, which may be said to have originated with Faraday's. Also, electromagnets, the telegraph, transformers, electric motors, and electric generators, all introduced shortly after Faraday's suggestion, followed logically from it. The work on transformers later led to the widespread distribution of electricity that we have today. Faraday himself made many significant experiments that led to important technical advances.

Faraday was unable to express his ideas mathematically, and this was done gradually by Clerk Maxwell over a period of a quarter of a century, from about 1854 until his death in 1879. Maxwell later insisted that he had done no more than put Faraday's ideas into mathematical form, but here he was being unduly modest. The fact that Maxwell, an excellent mathematician, had to work so long and hard over his theory tells us that the task was far from straightforward. Aside from that, Maxwell had to go into some aspects not considered by Faraday. For example, Maxwell investigated the speed with which an electric current passes along a wire, and found that it was the speed of light. This was a conclusion of great importance, because it demonstrated the connection between light, electricity, and magnetism.

It is impossible to explain Maxwell's theory in words, since everything is expressed in a few mathematical equations that have been described as "simple." They are indeed simple to look at, but understanding them requires a considerable background knowledge of electrical and magnetic theory, of quantities known as vectors, and of partial differential equations. One way of visualizing the theory in a simple way is to think of the vibrations shown in figure 13 (see page 79) as occurring in both the electric and the magnetic fields that exist around a wire carrying an electric current. Some of the consequences of the theory are easy to explain. One of these is that, besides light, there are many kinds of radiation that are invisible to the eye, and that it ought to be possible to produce and detect some of these by electrical means—something that had never been contemplated. A special kind of radiation of this kind was first produced in 1887, eight years after Maxwell's death at the age of forty-eight, by the German physicist Heinrich Rudolph Hertz (1857–1894). His apparatus was fairly simple. He found that a spark would generate electromagnetic waves having wavelengths of four to five meters, more than a million times greater than those in visible light. He was able to detect them electrically after they traveled across the room, by means of large coils with a gap in them, at which sparks were produced. You could say this was the first radio transmission, although much work had to be done to make it of practical use. Hertz always acknowledged that his success was only possible with Maxwell's theory. Later, Einstein said that his theory of relativity had been inspired by Maxwell's electromagnetic theory, of which he made great use.

A chart of the electromagnetic spectrum is shown in figure 16. The visible spectrum is only a small slice of the radiations known to us today. Before Hertz did his work, not much more than the visible spectrum was known. In 1800 the great astronomer William Herschel (1738–1822) discovered, through the heating effect of radiation, that there was some invisible radiation beyond the red, and this is known as the infrared. The wavelengths are longer and the frequencies lower. Einstein showed that light could be regarded as a stream of particles, later called photons. The photon energies (equal to the Planck constant multiplied by the frequency) are lower in the infrared than in visible light. In 1801 radiation beyond the violet, known as the *ultraviolet*, was discovered from the fact that it caused silver chloride to blacken. Here the photon energies are higher than for visible light.

By his experiments, done at wavelengths of four to five meters, Hertz greatly extended the electromagnetic spectrum by discovering the region of

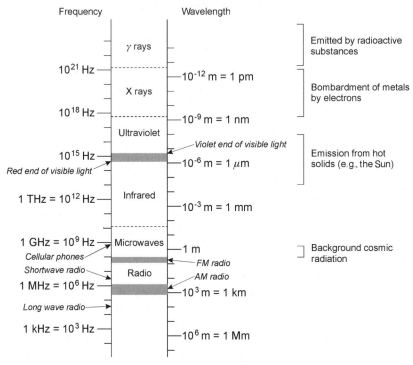

Fig. 16. The spectrum of electromagnetic radiation, showing some regions of special interest, and indicating how the various types of radiation are produced. Radiations at the top of the diagram (high frequencies and short wavelengths) are high energy radiations, while microwaves and radio waves at the bottom (long wavelengths and low frequencies) are very low energy.

the spectrum now called the *microwave region* (see figure 16). His wavelengths were more than a million times greater than previously observed. The wavelengths used today for radio and television are beyond the microwave region, in what is called the radio region. There are also some important radiations where the wavelengths are lower, and the photon energies therefore higher, than for ultraviolet radiation. For example, x-rays were discovered in 1895 by the German physicist Wilhelm Konrad Röntgen (1845–1923). Gamma rays (γ rays), which are emitted by radioactive substances and are present in cosmic rays, were discovered in 1900 by the French physicist Paul Ulrich Villard (1860–1934).

The important thing about electromagnetic radiation is that radiations at the top of figure 16 are high-energy radiations that penetrate matter more easily and bring about chemical reactions easily. This is why exposure of the skin to ultraviolet radiation, x-rays, or gamma rays can have undesirable effects. At the other end of the spectrum, low-energy rays such as radio waves cause no ill effects when they pass through our bodies, which, with so much radio and television broadcasting, they are doing all the time whether we like it or not.

Even after the quantum theory had been generally accepted by physicists, many scientists paid little attention to it. Chemists and biologists thought that they could get along nicely without it. All this changed in 1913, however, when the Danish physicist Niels Bohr (1885–1962; see figure 17) showed how the quantum theory can explain the structure of atoms and the nature of the chemical bond.

Bohr applied the quantum theory to the hydrogen atom, which consists of a positively charged nucleus and one electron (see figure 7 on page 63). He concluded that electrons are required to remain in fixed orbits around the nucleus. In the hydrogen atom the electron is in a particular orbit, but it can be "excited" to certain other orbits further from the nucleus. Only certain orbits are allowed by the quantum theory. Bohr worked out mathematical equations that expressed the positions and energies of the permissible orbits. The basic idea behind his interpretation of the spectrum of an atom can be understood by referring back to figure 14 (see page 81), which shows two of the orbits that are permitted according to the quantum theory. It requires energy to move an electron from the inner orbit to the outer one, because the electron is attracted by the nucleus. When light interacts with the atom that has an electron in the inner orbit, it is possible for a photon to be absorbed, provided that its energy is exactly that required to move the electron to the outer orbit. The photon, the energy of which corresponds to a certain wavelength of light, is annihilated. Examination of the *absorption spectrum* will thus show that light has been removed at this wavelength.

Conversely, when an *emission spectrum* is observed, an electron drops from an outer orbit to an inner one. A photon is emitted, with energy equal to hn, which is exactly equal to the difference between the energies of the two levels. Bohr applied his theory to the spectrum of the hydrogen atom, and found remarkably good agreement with the experimental results.

Bohr's theory was later superseded by *quantum mechanics*, but its basic ideas still remain of great value to chemists since they provide a convenient way of visualizing how even quite complicated molecules are put together. In other words, the theory provides a simple way of thinking about chemical bonds. In 1916 the American chemist Gilbert Newton Lewis (1875–1946) made a contribution of special significance by suggesting that the type of chemical bond in simple molecules involves a pair of electrons. His idea was that bonding can take place when two atoms share electrons. His original suggestion, later modified, was that groups of eight electrons ("octets") were stationed at the corners of cubes. Bonding

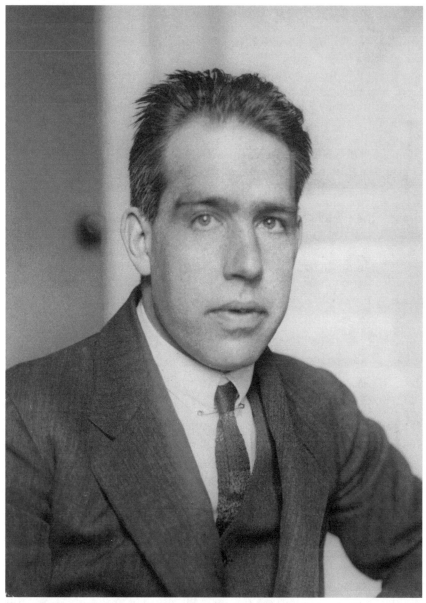

Fig. 17. Niels Bohr is particularly distinguished for his electronic theory of the atom. He also did important work on the structure of atomic nuclei, particularly in connection with nuclear processes. Throughout his career he exerted a strong influence on the development of atomic and nuclear physics, not only through his own work but by his support of others. (Courtesy of the Edgar Fahs Smith Collection, Annenberg Rare Book and Manuscript Library, University of Pennsylvania.)

could occur by an overlap of the edges of two. He recognized that in some molecules both of the bonding electrons may originally have come from one of the atoms. With the advent of wave mechanics, Lewis's ideas were greatly developed, and in modified form are still used today by chemists as a valuable way of visualizing the electronic structures of molecules.

There is yet another universal principle that we need to know, if we are to understand more of nature's mysteries. We have seen that radiation has particle properties as well as wave properties. The converse idea, that particles can have wave properties, was proposed by Louis Victor, Prince de Broglie (1892–1987). He suggested that particles such as electrons, when moving along in a beam, can behave as if they are waves. By making use of radiation theory, he deduced that the wavelength λ of a moving particle is equal to the Planck constant h divided by mv, its mass m multiplied by its speed v. The product of the mass and the velocity, mv, is called the particle's momentum, p, so the wavelength λ of the particle is the Planck constant divided by the momentum:

$$\lambda = \frac{h}{mv} = \frac{h}{p}$$

A consequence of this is that a beam of electrons can be diffracted, just like a light beam, and experiments involving electron diffraction are often carried out today.

This suggestion inspired the Austrian physicist Erwin Schrödinger (1887–1961) to formulate a new mathematical treatment, called *wave mechanics*, based on the quantum theory. In Schrödinger's wave mechanics, an electron in an atom is treated as a wave instead of as a particle orbiting around the nucleus, which was the case in Bohr's theory. Schrödinger did not arrive at his equation by a formal mathematical proof, but instead proceeded by analogy with the electromagnetic equations of Maxwell. In his theory, quantum restrictions are not introduced arbitrarily, but appear as a direct mathematical consequence of the wave equation. No solution of this is possible unless the energy has one of a number of permitted values. An electron's orbit restrictions result from the need for a wave to fit into the right amount of space. When we depict an atom on the basis of wave mechanics, we do not show a diagram like figure 14 (see page 81) in which the electrons move in definite orbits; instead, we show the electron as a kind of cloud hovering around the nucleus, and we use the word *orbital* rather than orbit. The density of the cloud is interpreted as representing the probability that the electron is in a particular position.

Several other formulations of wave mechanics, or quantum mechanics as it is also called, were made at about the same time, notably by the

German physicist Werner Heisenberg (1901–1976) and the English physicist Paul Adrien Maurice Dirac (1902–1984). These treatments were not as easy to visualize, and it was soon found that all of the treatments are mathematically equivalent. They have played an important role in modern physics and chemistry, and have led to many discoveries. For example, Dirac was able to deduce that an electron must have a *spin*. There are just two possible quantum numbers for the spin of an electron. The electron can rotate in one direction with a certain speed, and in the opposite direction with the same speed, but in no other way. Dirac also predicted the existence of an elementary particle having the mass of the electron but a positive charge instead of a negative one. This particle, called a *positron*, was discovered in 1932 by the American physicist Carl David Anderson (1905–1991).

One of the consequences of the wave theory of matter is the *uncertainty principle*, or *principle of indeterminacy*. Heisenberg arrived at the principle by carrying out an imaginary experiment in which a beam of light is used to determine the position and momentum (mass multiplied by velocity) of an electron. Because the light disturbs the electron, there is a fundamental and inescapable limitation on the accuracy of the measurements. His conclusion was that it is *impossible to determine both the position and speed of a particle exactly*. The more accurately one knows the position, the less accurately one knows its speed, and vice versa.

This principle is important not only in physics but in philosophy. Previously, people had argued that if we knew the exact state of the universe at any one time, the future would be completely determined. That would imply that humans would not have free will, since our behavior would be completely determined by our physical make-up. The principle of indeterminacy shows that this is not the case, because our exact state cannot be determined, even in principle. There are constraints on our actions, but our free will allows us to make choices as to how we behave.

Such ideas were first put forward by Niels Bohr and are often referred to by scientists as the *Copenhagen interpretation*, since Bohr was a professor at the University of Copenhagen. The essential feature of the Copenhagen interpretation is that there is a lack of complete determinism, in that future events do not follow inevitably from past conditions; pure chance plays a role. Einstein and Schrödinger both took strong exception to this idea, and scientists still disagree about whether the interpretation is acceptable. The question is more a philosophical than a scientific one, and for that reason is impossible to settle. I personally am quite satisfied with the Copenhagen interpretation, which seems to me to give the simplest explanation of all the experimental evidence.

Einstein's objection to the Copenhagen interpretation is summarized in his often quoted statement that "God does not play dice" in connection with the events of the universe. This comment needs a little interpretation, since Einstein was always at pains to explain that he did not believe in God in any conventional sense. What Einstein meant was that he did not believe that the universe is constructed in a way that would allow events to occur entirely by chance. Today, however, most physicists have accepted that chance does play a role. As far as we are able to determine, individual radioactive transformations occur as a matter of chance; we cannot predict when a particular radioactive nucleus is going to emit radiation, because Heisenberg's principle of indeterminacy says that it is impossible to make sufficiently precise measurements.

There is another and completely different reason why it is impossible to make predictions of future events. This is the subject of the theory of *chaos*, which we hear a lot about today, but unfortunately, it is much misunderstood. This is partly because scientists are now using the word chaos in two different senses that are somewhat related, but must be clearly distinguished. The first and original sense corresponds to the dictionary definition of chaos, "utter confusion." This is the meaning that we use in connection with the second law of thermodynamics. Processes occur in the direction of increasing entropy—increasing disorder—and we speak of an "arrow of time" corresponding to increasing disorder and leading eventually to chaos. In other words, the universe is becoming more and more disordered, so eventually there will be complete chaos. Of course, it will be many billions of years before anything like this happens.

The type of chaos referred to in the modern theory of disorder is much more limited.[4] It denotes certain occurrences for which the final outcome is impossible to predict for reasons *other* than the principle of uncertainty. Over the past few decades, a great deal of experimental and mathematical investigation has shown that unpredictable behavior is much more common than had previously been supposed, and that it has a perfectly rational explanation. Mathematical and computational work has led to what is generally, but regrettably, called *chaos* theory and sometimes—even less appropriately—*catastrophe* theory. This kind of chaos is obviously very different from the complete chaos that will exist at the end of time, and it is unfortunate that the same word has come to be used for it. Chaos in this much more limited sense is sometimes referred to as *noncatastrophic chaos*. In this chapter we will follow the usual convention of simply calling it chaos.

Consider a familiar process, the bouncing of a tennis ball from a racket. With practice, we can hold a racket horizontally and control the up-and-down motion of the bouncing ball so that the ball always rises to the same height. Suppose, on the other hand, that we hit the ball with a predetermined force and a frequency that is independent of the motion of the ball. The ball will then bounce up to a different height each time. After a few bounces the height of a bounce will be quite unpredictable, even if we measure the conditions as precisely as possible. This is chaos, in the modern scientific sense of the word, although it is chaos in a limited sense—noncatastrophic chaos—since obviously the rest of the universe is hardly affected.

There are many other familiar examples of this kind of chaos. If we are pushing a child in a swing, we normally avoid the possibility of chaos by coordinating our pushing to the natural motion of the swing. If instead we perversely pushed at a fixed frequency, and without such coordination, the swing would perform a chaotic motion (anyone trying the experiment should not have a child in the swing). Our lives frequently involve avoiding chaos of this kind. If we steer and brake a car in an irresponsible manner, without regard to pedestrians or traffic, we will soon achieve chaos—and perhaps a court appearance.

Chaos is sometimes the outcome of oscillatory behavior. An example is a heartbeat, which is controlled by natural pacemakers. Sometimes these do not work together properly, so that there are alternate long and short gaps between the beats. Under more extreme conditions the beating becomes highly irregular. In one particularly dangerous condition, called *ventricular fibrillation*, the heart flutters erratically instead of expanding and contracting rhythmically. A small change in the timing of one beat makes a bigger change in the timing of the next; the beating becomes chaotic. One feature of ventricular fibrillation that makes it difficult to treat is that the individual processes in the heart may be working normally; it is their lack of coordination that leads to overall malfunctioning.

Up until the nineteenth century, the prevailing scientific theories were based on a *deterministic* view, the assumption that the future is entirely determined by the past. This had been assumed by Isaac Newton, and a century later, Pierre Simon Laplace (1749–1827) argued that if we could specify the exact state of the universe at a given time, we could in principle deduce its state at any future time. The first challenge to that idea came from the French mathematician and physicist Jules Henri Poincaré (1854–1912). In 1909 he pointed out that there can be situations where small differences in the initial conditions can produce very large ones in the final result, so that predictions become impossible. Although this is a very clear statement of the conclusions of modern chaos theory, few people at the time paid any attention to it.

Another kind of challenge to determinism came in the 1920s with the advent of quantum mechanics and in particular with Heisenberg's formulation of the uncertainty principle. According to this principle, it is impossible to determine, at the same time, the exact position and momentum of a particle. As a result, we cannot predict the exact course of events, but can sometimes make reliable estimates of probabilities.

Chaos theory, however, relates to a *deterministic* type of uncertainty, which would occur even if there were no restriction imposed by the Heisenberg principle. This type of chaos is sometimes called *deterministic chaos*. Even if the uncertainty principle did not affect the situation, we still could not always predict what is going to occur. When one analyzes certain kinds of dynamical situations mathematically, there is a basic instability in the behavior. For example, two computer calculations carried out with very tiny differences in starting conditions can lead to entirely different outcomes.

One of the first important mathematical treatments leading to modern chaos theory was put forward in the 1960s by the American meteorologist Edward Lorenz (b. 1917) of the Massachusetts Institute of Technology. He used a computer to study weather patterns, and devised highly simplified differential equations to represent meteorological processes such as the movement of air and the evaporation of water. His aim was to make predictions for the temperature, the direction of the wind, and the onset of rain or snow. He introduced into his programs certain initial conditions and then let the computer predict the values of the various meteorological parameters after various periods of time. To his great surprise he found, even with his highly simplified model, that when he repeated the calculations with almost exactly the same initial conditions, the predictions might be much the same for the first few days, but after that there were enormous divergences, and no similarity between the predictions remained after more than a few days.

For example, he made one set of calculations putting in a number 0.506127, and then did the same calculations rounding off this number to 0.506, assuming that the difference of one part in 5,000 would make no difference. One part in 5,000 might correspond to a tiny movement of air or a temperature difference that could hardly be measured. However, this tiny difference led to completely different predictions about the weather a few days later. At first Lorenz assumed that the computer was malfunctioning, but further work showed the effect to be real.

This means that there are fundamental limits to our ability to predict the weather. If predictions are highly unreliable even when the mathematical equations have been greatly simplified, they will be more unreliable for the conditions actually existing. Experts can fairly reliably predict the weather for the next two or three days. After a few days, however, computers will make a wide range of predictions from exactly the same starting conditions, which means that predictions are really impossible. In 1972 Lorenz suggested what is called the "butterfly effect," with his paper "Does the flap of a butterfly's wings in Brazil set off a tornado in Texas?"

Another notable contribution relating to chaos theory was made in the early 1970s by the Polish-French mathematical physicist Benoit Mandelbrot (b. 1924). One of the by-products of his work was his construction of special patterns and his introduction of the idea of fractals. The importance of these is that they show that complex structures can easily arise from highly disordered systems.

It is rather a paradox that, although chaos theory has developed so recently, it deals with many very familiar things: water dripping from a leaky tap, a flag flapping in the wind, traffic jams on the highway, and the rise and fall of the stock market. It is also relevant to such common occurrences as a political protest, which may start with everyone behaving in a reasonable manner, but ends in a violent riot. The theory turns everyday happenings, previously avoided and sometimes scorned by scientists, into legitimate subjects of scientific study. Chaos theory unites research workers in different fields and works against the excessive specialization that is so prevalent today.

There are exact conditions under which chaos develops. One condition is called *positive feedback*. We can best understand positive feedback by first considering negative feedback, an example being the thermostats that control the temperature in our homes. If the temperature gets too high, the furnace is turned off; then, when the temperature gets too low, the heat turns on again. Suppose that our thermostat, instead of working in that way, has been wired by a maverick electrician so that if the temperature is too low the furnace goes off, and if it is too high the furnace goes on. If the temperature is too high, the furnace makes it still higher, and the temperature eventually settles down at some high value, corresponding to the best that the furnace can do. That is positive feedback. Positive feedback happens in a chemical or nuclear explosion; a nuclear explosion, for example, occurs when uranium-235 reaches a critical mass and the chain reaction results.

A familiar example of positive feedback happens at an event such as an office party. Often, the noise soon becomes so loud that we have to shout at the tops of our voices. It would obviously make more sense if everyone agreed to speak in a normal voice, but that would only work if police were assigned to haul away offenders, and that is probably not the most popular way to give a party. If there were such a prior agreement, but no police, we can be sure that someone would soon speak a little louder, forcing others to do the same, and the sound would soon reach a high level.

Here is another example of chaos in this more limited sense. Suppose that we have a lawn that becomes infested with beetles, and, to keep things simple, suppose we do nothing about it, merely watching in dismay. The lawn starts to deteriorate and the beetle colony grows exponentially. The lawn becomes so bad that beetles die of starvation and only a few remain. The lawn then revives and begins to look acceptable; then the few beetles still alive begin to replicate, and the cycle starts again. It is easy to see that the quality of the lawn may go through maxima and minima, the population of beetles doing the same. There is obviously feedback in this case. The number of beetles affects (adversely) the growth of the grass, while the amount of grass affects (favorably) the growth of the beetle colony; we have negative feedback in one direction and positive in the other, and it is the positive feedback that is essential to chaos.

What will be the final outcome? One possibility is that the grass and the beetles might all die. Or the lawn might settle into an equilibrium state where the beetle population remains fairly low and the lawn is not too bad but not at its best. What would eventually happen is impossible to predict, even if there were no outside influences. This is true even if we ignore environmental factors; the outcome would be affected by the weather, for example.

Here is yet another situation in which chaos relates to everyday life. Suppose that you were playing billiards and wanted to make a cannon— that is, to hit two balls in succession. If I were standing nearby, would you make an adjustment for the gravitational force that I was exerting? Not very likely. Suppose, on the other hand, that you were ambitious enough to try to hit nine balls in succession; should you then take my gravity into account? The answer to this, according to French mathematician Ivar Ekeland's fascinating book *Mathematics and the Unexpected*,[5] is yes, you should. Of course, in practice it is quite impossible. But in principle, that tiny gravitational force affects the outcome of a nine-ball cannon, because of the multiplication of the effects when so many collisions have taken place. Even the gravitational force of an electron at the far outer edge of the universe, 10^{10} light-years away, will make a difference after a cannon of enough balls—about fifty-six, according to the calculations.

Of course, there are many familiar processes that do not give rise to chaos. Examples are the swinging of a simple pendulum, ordinary engines like those in automobiles, and most chemical reactions. For these, with the aid of the appropriate theory, we can predict exactly and without any ambiguity just how the process will occur over a period of time. The behavior may be complex and difficult to work out, and we may have to use a computer, but the processes present no surprises. What occurs is just what we expect. This is fortunate in many practical situations; it would be disastrous if, when we applied the brake to a car, it sometimes accelerated for no ascertainable reason.

Even if there is the possibility of chaos in a particular situation, it is sometimes possible to control it. We saw one possibility of this, in the case of a person controlling the motion of a child's swing. Similarly, chaos on the stock exchange can sometimes be controlled by the skillful action of a financial authority. To control our personal lives we constantly have to look out for impending chaos, and try to head it off. This the kind of thing that William Shakespeare no doubt had in mind when he wrote in *Julius Caesar*:

There is a tide in the affairs of men,
Which, taken at the flood, leads on to fortune;
Omitted, all the voyage of their life
Is bound in shallows and in miseries.

Chapter 4
Our Place in the
Universe

It is the stars,
The stars above us, govern our condition.

—William Shakespeare, *King Lear*, Act 4, Scene 3

We used to think our fate was in the stars. Now we know, in large measure, our fate is in our genes.

—James D. Watson, quoted in *Time*, March 20, 1989

Astronomy, geology, and biology are the sciences that are concerned with, among other things, human beings and our place in the universe. To understand ourselves and other forms of life, we turn to biology. To get some idea of how life developed on Earth we need additional help from the science of geology. To know how Earth and, indeed, our universe might have come about, we look to the stars—astronomy.

Parts of these three fields of science are soft, in the sense that they are not yet as precise, and cannot be treated as mathematically, as physics and chemistry. Until about the middle of the nineteenth century, all three were mainly concerned with the collection of data, just as physics and chemistry were in their earlier years. More recently, astronomy, geology, and biology have all developed very hard areas. What is particularly striking is that these sciences come to much the same conclusions about the origin of Earth and of life on Earth.

Because these sciences remain softer than physics and chemistry, the evidence on which some theories are based is often more circumstantial. Sometimes detractors of science regard this as a weakness, but this is not really the case. As we discussed earlier, circumstantial evidence is often stronger and more reliable than direct evidence: as the American writer Henry David Thoreau put it, "Some circumstantial evidence is very strong, as when you find a trout in the milk."[1] Even in the hard sciences, many important theories, such as the theory of relativity, were initially deduced from circumstantial evidence. At the beginning of the twentieth century hardly anyone doubted the reality of atoms, though they had been deduced to exist only on the basis of quite indirect evidence; today the evidence is more direct. We shall see a particularly interesting example of powerful circumstantial evidence in the next chapter, where quite independent results from astronomy, physics, chemistry, geology, and biology all lead inexorably to the conclusion that Earth must be more than four billion years old. Some people still doubt this conclusion, but they should give careful consideration to the large amount of observational evidence from widely different branches of science.

Theories in the soft sciences tend to be criticized by nonscientists more than theories in physics and chemistry. The main reason is that the hard sciences are not considered to have much impact on philosophical and religious opinions. It is true, as we have seen, that Einstein did have some trouble with philosophers after proposing his theory of relativity, but that was unusual. Astronomers, geologists, and biologists, on the other hand, are quite often criticized by people who have little understanding of science, but who nevertheless express their opinions freely on scientific work.

Aside from this problem of uninformed criticism, it must also be admitted that scientists themselves have at times been less than reasonable, rejecting valid observational and experimental evidence that does not agree with their preconceived ideas. Sometimes their objections had a religious basis, perhaps concealed by a blanket of spurious scientific arguments. In the eighteenth century, the geologist James Hutton presented overwhelming evidence for the great age of Earth, but some geologists tried to counter it with misinterpretations and false evidence, possibly without realizing that they were the victims of ingrained religious prejudices. Similar unreasonable objections were also raised against Louis Agassiz and his theory of glaciers, Charles Darwin and his theory of evolution, and Alfred Wegener and his theory of continental drift.

Aside from difficulties of this kind, scientific fraud has occasionally occurred in the soft sciences. A well-known example is the so-called Piltdown man.[2] In 1912 Charles Dawson, a lawyer who was an avid collector of fossils, exhibited a skull that appeared to be that of a man who lived at the beginning of the Glacial or Pleistocene era—nearly a million years ago. With the skull were flint tools and a pointed bone tool. He claimed to have made his discovery at Piltdown, which is near Lewes in Sussex, England. This discovery was greeted with great excitement, since it seemed that the skull was a "missing link" between ape and man, but it did not fit in with theories about human evolution. Detailed technical examination made in the 1950s revealed that the whole thing was a hoax, probably perpetrated by Dawson himself. The jaw was that of an orangutan, and the cranium, although human, had been stained to increase its apparent age. With the development of modern analytical techniques it is now impossible for anyone to get away with this kind of hoax.

As in the hard sciences, some errors arise as a result of genuine mistakes, usually associated with self-delusion. The idea of canals on the planet Mars began with the distinguished Italian astronomer Giovanni Virginio Schiarparelli, who in the 1860s became head of the observatory and museum in Brera near Milan. He produced excellent maps of the surface of Mars and reported that he had detected a number of straight lines and curves, which he sometimes called *canali* (channels) and sometimes *flume* (rivers). The word *canali* suggests the English word canals, which led some people to think that they were built by living creatures on Mars. This idea, which had never been taken seriously by Schiarparelli himself, was taken up with great enthusiasm by Percival Lowell (1853–1916), a wealthy amateur who established an excellent observatory in Arizona and published a great deal on the subject of Mars and its "canals." He believed that, because of the parched terrain of Mars, the Martians built canals for irrigation. For decades it was widely believed that there existed a race of Martians who, besides building canals, sent out radio signals. The theme of envious and thirsty Martians turning their eyes toward our green Earth was cleverly exploited by H. G. Wells in his novel *War of the Worlds*, published in 1898. As late as 1938, a radio program designed by Orson Welles and based on this novel created widespread panic for a few hours in the United States, when many Americans became convinced that the Martians had finally invaded.[3]

Most scientists remained skeptical about the idea that living beings on Mars had built canals. As technologies improved, the hopes of finding life on Mars dwindled. It is now known that Mars is most inhospitable to any higher form of life on account of its lack of surface water, its very low temperatures (down to $-160°C$), and its atmosphere—mainly carbon dioxide at a tiny pressure. There is evidence, however, that at one time Mars had lakes and rivers, and an atmosphere more like our own.

*   *   *

It is useful to glance at the history of astronomy to see how it has developed from a very soft science to one that encompasses many hard areas.

Astronomy is the oldest of the numerical sciences, beginning in about 3000 BCE, when people, particularly in the Near East and China, began to group the stars into constellations. Monuments, such as the pyramids in Egypt and Stonehenge in England, provide evidence for an early interest in astronomy. The relative movements of the Sun and Moon led to a primitive form of calendar defining the year, the month, and the week. The first satisfactory proof that the earth is spherical was given by Aristotle (384–322 BCE), who, because of his great authority, created much controversy by his insistence that the earth was the center of the universe. Aristarchus (c. 320–c. 250 BCE), one of the most accurate and original of the Greek astronomers, attempted to estimate the distances from Earth to the Sun and the Moon, but his results were rather inaccurate. He was probably the first to propose that it is better to regard Earth as going around the Sun, rather than the other way around. Another famous Greek astronomer was Hipparchus of Rhodes (c. 170–c.125 BCE), who constructed a detailed catalogue of 850 stars and constructed a scale of magnitudes to indicate their relative luminosity or brightness. He is commonly regarded as the founder of trigonometry, a valuable mathematical tool.

The Egyptian-Greek astronomer Claudius Ptolemaeus (usually known as Ptolemy, c. 90–170 CE) made a detailed study of the movement of the Sun and the planets, and proposed a detailed mathematical model in which Earth was at the center and the Sun and planets moved around it in orbits. This Ptolemaic universe was accepted for about thirteen centuries, and little progress was made in astronomy or indeed in any branch of science. The main reason was that, for many centuries, scientific procedures became what we should today call "politically incorrect," and were supplanted by less mentally challenging theological and astrological studies.

One of the first steps out of this intellectual darkness was taken by the Polish priest Nicolaus Copernicus (1473–1543), who worked mainly from the astronomical data of others rather than from his own observations. He recognized that in explaining the movements of the planets it was more satisfactory to assume that they revolve around the Sun. He first expressed his ideas privately in 1514 in a short manuscript, and continued to develop them for the next thirty years. He explained the apparent movement of the Sun around Earth as due to the rotation of Earth. He also suggested that the stars are much farther away than had previously been proposed. He expounded his ideas in *De revolutionibus orbium coelestium* (*The Revolution of the Heavenly Spheres*), published in the year of his death; he may have seen a copy of the published book only on the day he died. Copernicus's work was one of the most important and original advances in astronomy.

Much more astronomical information was required to establish Copernicus's ideas. Important work was done by the Danish astronomer Tycho Brahe (1546–1661), who greatly improved the accuracy of astronomical measurements and made a vast number of observations, measuring the positions of 777 stars with great accuracy. Brahe never became a Copernican, but took the compromise position that all of the planets except Earth go around the Sun, which itself goes around Earth. It is worth noting in this connection that, in fact, the Sun does go around Earth; each goes around the other. The important point, as Copernicus had concluded, is that we get a mathematically simpler concept of the entire solar system by regarding all the planets as going around the Sun. It took many years to rule out Brahe's compromise, which led to predictions that were similar to those of Copernicus.

In 1600 the German astronomer Johannes Kepler (1571–1630) went to work with Brahe, and there is a story that on his deathbed Brahe begged Kepler not to use his observations to support the Copernican hypothesis. Kepler's great contributions were to discover that planets have elliptical rather than circular orbits, and to establish empirical laws describing their motion. His work tended to support the Copernican ideas, but Kepler placed more stress on the shapes of the orbits rather than on whether the Sun or the earth was the center of the system.

Galileo Galilei (1564–1642) developed a greatly improved telescope and made a series of spectacular discoveries. Between 1609 and 1610 he made observations on four satellites of Jupiter and on the orbit of Venus. All his work led him to accept the Copernican model for the solar system. The Roman Catholic Church ordered Galileo not to follow the Copernican theory. An interesting detail is that Galileo sold his telescope to the Venetian republic for commercial and military use; this is an early example of the connection that often exists between pure science and practical applications.

Galileo also made important contributions to the science of mechanics. In particular, he established, by sliding weights down inclined planes, that their motion was independent of their mass; previously it had been assumed that a heavy weight would fall faster than a light one. The story that he dropped weights from the top of the Leaning Tower of Pisa is a nice one, but unfortunately it appears to be untrue. It is interesting that Newton, in his *Principia Mathematica*, later described experiments, carried out in 1710, in which bodies were dropped from the cupola of St. Paul's Cathedral in London and measurements were made of the times it took them to land.[4] It is curious that these experiments are today rarely mentioned, whereas almost everyone "knows" the apparently untrue story about Galileo and the tower of Pisa. In other experiments, Galileo established the relationship between the force on a body and its acceleration. However, he was not able to relate his mechanics to his astronomical observations on the orbits of planets; that was done by Newton.

Francis Bacon, whom we met earlier, was almost contemporary with

Galileo. Bacon emphasized that general principles must always be tested against experimental results. His writings exerted great influence, and Newton followed the path laid out by Bacon. Isaac Newton was born in the year of Galileo's death, and was still in his early twenties at Trinity College, Cambridge, when he clarified Galileo's mechanical principles, stating his three laws of motion. All of his studies gave great support to the Copernican theory, and by interpreting the motion of the planets in terms of his gravitational law and of his mechanical laws, he united celestial and terrestrial science for the first time. Newton also did mathematical work on the motion of planets, work that was followed by a similar treatment by Edmund Halley (1656–1742), famous for the comet bearing his name. This came close to Earth in 1682, and Halley predicted that it would do so again in 1758, which it did.

Newton's mathematical treatment of the solar system was one of the greatest achievements in astronomy. It led to many other investigations, such as those of William Herschel (1738–1822) who interpreted binary stars in accordance with Newton's theory. He also studied the Milky Way, that remarkable band of light that stretches across the night sky. Prior to Herschel, the suggestion had been made by the English instrument maker, Thomas Wright (1711–1786), that the Milky Way is part of our galaxy, which has the shape of a flat disk; he was correct. Before Herschel, little else was known about the Milky Way. Herschel constructed a number of telescopes, making larger and larger ones, and observed a number of nebulae, which have a different appearance from the stars in the Milky Way. He found that by increasing the size of his telescopes and resolving the stars in some of them, he could observe many more nebulae in which he could not distinguish individual stars. He correctly thought, but could not prove, that many of these nebulae lay outside our galaxy and were star systems like the Milky Way. These were later called *extragalactic nebulae*. Some of them, including our own, have spiral structures and are known as *spiral nebulae*. Today we usually call them all galaxies.

Herschel, like all astronomers in his time, could not measure distances from Earth to even the nearest stars, let alone the nebulae. Before we go into how this was later done, we should consider a common unit for measuring great distances, the *light-year*, which is the distance that light travels in a year. Light travels at about 300,000,000 meters (186,420 miles) a second, or about 300,000 kilometers a second, and a year is roughly 32,000,000 seconds; a light-year is thus nearly 10,000,000,000,000 (10 million million, or 10^{13}) kilometers (about 6 million million miles). It is difficult to appreciate these enormous numbers and speeds, but consider this. Light takes about a hundredth of a second to cross the Atlantic, and about 1.3 seconds to get from here to the Moon, which averages about 384,000 kilometers (238,600 miles) from the earth. We can say that the

Moon is roughly 1.3 light-seconds away from us. Light reaches the Sun in about eight minutes, so the Sun is eight light-minutes from us. All of the stars that we see in the sky are more than four light-years away, and some are several billion light-years from us. Astronomers usually measure distances, not in light-years, but in parsecs; one parsec is equal to 3.262 light-years, or 3.086×10^{13} kilometers (18.6 million million miles).

Proxima Centauri is the star nearest to us, about 4.3 light-years away. To appreciate such great distances, suppose that a spacecraft existed that could travel at an average speed of ten kilometers a second, or 36,000 kilometers an hour (22,370 mph). This is about Mach 30, that is, about thirty times the speed of sound (1,200 km per hour or 746 mph). Such a high average speed has never been attained, but it might be a possibility in the future if nuclear-powered propulsion is employed. A spacecraft traveling at that rate would get us to the Moon in a little over ten hours. The speed of the spacecraft, although enormous, is nevertheless only a tiny fraction, 1/30,000, of the speed of light, and to travel a light-year would take about 30,000 years. To get to Proxima Centauri would take about 130,000 years. Proxima Centauri is not known to have any planets, and the nearest star that seems to have a planet is about twenty light-years away; to get to that would take about 600,000 years.

These facts are rather embarrassing to people who think that we might one day be invaded by creatures who occupy other planets, or that we might be able to colonize other regions of the universe. Life cannot be maintained on anything but a planet, since everywhere else is too hot or too cold. The other planets in our own solar system are inhospitable to any form of life, so we would have to contemplate traveling at least twenty light-years. However, a trip of 600,000 years is surely too much to contemplate, even by the most intrepid of travelers. Travel to or from a planet in any of the other galaxies would take much longer than the present age of the universe.

Stellar distances were first measured satisfactorily and almost simultaneously by German astronomer Friedrich Wilhelm Bessel (1784–1846), Scottish astronomer Thomas Henderson (1798–1844), and German-born Russian astronomer Friedrich Georg Wilhelm Struve (1793–1864). They used the parallax method, similar to the method used today by surveyors: by observing an object from two different positions, its distance from us can be calculated. All stars and galaxies are too far away for reliable measurements to be made from two positions on Earth, since they cannot be far enough apart. Copernicus had looked for stellar parallax, using observation points some distance apart on Earth, but had concluded that the stars were so far distant that the parallax was immeasurably small. However, in its orbital motion around the Sun, Earth moves considerable distances. Observations made at two different times, when the distance between positions is accurately known, provide a much longer base for the parallax measurements. This has made it possible to obtain reliable values for the distances of stars that are nearest to Earth. Today, much more accurate parallax measurements are being made from artificial satellites, such as the European satellite *Hipparcos*. Methods other than those using parallax must be used for all the nebulae (galaxies) and for the more distant stars.

In 1838, Bessel was the first person to announce a star's distance measured by the parallax method. He made his first measurement on the binary star 61 Cygni, and found its distance to be 10.3 light-years. The latest precise measurements made from the satellite *Hipparcos* are 11.36 light-years for the brighter component and 11.43 light-years for the fainter one. Bessel's value was thus too low, but nevertheless a great achievement in its time.

Thomas Henderson, who was director of the Royal Observatory at the Cape of Good Hope and later Astronomer Royal for Scotland, and Wilhelm Struve, member of a remarkable family of astronomers, had made similar observations even earlier than Bessel, but published them a little later because of their reluctance to publish before their results had been carefully checked.

Of particular importance for astronomy was a discovery in physics, made in 1842 by the Austrian physicist Christian Johann Doppler (1803–1853), professor of mathematics at the Realschule in Prague. His work was with sound waves, and he found that if sound of a certain pitch or frequency is emitted by an approaching source, it appears to have a higher pitch than if the source is stationary. If the source is moving away, the pitch is lower. A convincing test of the Doppler effect was made in 1845 in the countryside near Utrecht and at the suggestion of the Dutch meteorologist Christoph Heinrich Dietrich Buys Ballot (1817–1890). An open railway carriage containing trumpeters was pulled past a group of musicians, who confirmed that as the train approached them the pitch was indeed higher, and that it became lower after the carriage had passed them. Today we have many opportunities of appreciating the Doppler effect, at airports for example, or when police sirens pass us at high speed.

Doppler suggested the same effect for light waves. If a star is receding from us, the light it emits will be of a lower frequency than if the source were at rest. Since the lower frequency (longer wavelength) end of the visible spectrum is the red end, the lines in the visible spectrum of the star shift toward the red end, and as a result we often speak of the *red shift*. It is important to realize, however, that the Doppler effect has nothing to do with the color of stars. It is true that the blue light from a receding star shifts toward the red, but some of the star's ultraviolet light shifts into the blue part of the spectrum, so the observed color hardly changes.

Doppler himself thought that the overall color changed, but Buys Ballot pointed out his error. Early estimates of the speeds of stars from their colors led to unreasonably large values, and it was realized that other factors, particularly the age of a star, influence its color as well.

Once this point was clarified and attention focused on the individual lines of the stellar spectra, the Doppler effect became a powerful tool in astronomical spectroscopy. The credit for recognizing this goes to Armand Hippolyte Louis Fizeau (1819–1896), a French physicist of remarkable versatility, and the first to measure accurately the speed of light. In the spectrum of a receding galaxy every line has a lower frequency, which means a longer wavelength, than when we observe the spectrum of the same substance in the laboratory.

Our understanding of the nature of the universe was greatly clarified by the contributions of the distinguished American astronomer Edwin Powell Hubble (1889–1953; see figure 18), who did most of his work at the Mount Wilson Observatory in California. He studied many of the nebulae and was particularly concerned with their distances from Earth. All of them, he found, were considerably farther away than even the most remote regions of the Milky Way. His work proved that spiral nebulae are distant galaxies, referred to as extragalactic galaxies. The nearest galaxy to us, Andromeda (M31), is two million light-years away, which is ten times farther than the most distant stars in our own galaxy. Using the latest techniques, astronomers can see galaxies that are over ten billion light-years away. The light we see from a galaxy that is ten billion light-years away was emitted ten billion years ago; anything we can learn about such a galaxy is very ancient history. Also, the place where we see it is actually where it was ten billion years ago. If the universe is twelve billion years old, that galaxy appears to be where it was two billion years after the universe was formed.[5] The fact that there are so many galaxies at such great distances is strong support for the great age of the universe and its enormous size.

The most recent work in astronomy has established that a galaxy is a system of stars, gas, and dust held in one region of space by gravitational attraction. Rather than being spread around randomly in space, galaxies are mostly in clusters, pulled together by gravity. Our own galaxy is relatively close to Andromeda and not far from twenty or so smaller galaxies; together these galaxies are known as the Local Group. The visible part of our galaxy consists of a flat disk of stars, with a diameter of about 80,000 light-years and a thickness of 6,000 light-years; it has enormous spiral arms. Our Sun and its planets, comprising the solar system, are about 30,000 light-years from the center of the disk. If we look out along the plane of the flat disk from Earth we see many more stars than in other directions. This large group of stars constitutes the Milky Way.

Radio waves are also received from outer space, and their investigation has revealed many new celestial objects. Figure 16 on page 86 shows just where radio waves are, relative to other parts of the spectrum. Radio waves

Fig. 18. American astronomer Edwin Powell Hubble found that spiral nebulae, such as Andromeda, are independent galaxies. He concluded that galaxies recede from us with speeds that increase with their distance. He also made estimates of the size and age of the universe. (Courtesy of the Archival Services, California Institute of Technology.)

and microwaves are much less energetic than the waves in visible radiation; the photons in them carry about a million times less energy than the visible. Astronomer Sir Martin Ryle (1918–1984) had a striking way of calling attention to the smallness of the energy when members of the public came to his observatory outside Cambridge. They were asked to pick up a piece of paper as they left, and on it was written "On picking this up you have expended more energy than has been received by all the world's radio telescopes since they were built." In the early seventies all the world's radio telescopes had not yet received enough energy to strike a match.

Radio waves from space were first investigated in 1933 by American radio engineer Karl G. Jansky (1905–1950) of Bell Telephone Laboratories in New Jersey. His work was at radio frequencies of 20.3 megahertz (MHz), which corresponds to a wavelength of about fifteen meters, used in short-wave radio transmission. He detected some background radiation, which he proved did not come from thunderstorms or from the Sun, and he finally established that it was strongest in the center of our galaxy. The announcement of this discovery made the front page of the *New York Times*, but Jansky did little further work in the field, and surprisingly, his results were largely ignored by astronomers.

Grote Reber (1911–2002) was one of the few who did pay attention to Jansky's observations. He was an American radio engineer, and in 1937, at his own expense, he built the world's first radio telescope in his backyard. For the next few years he was the only radio astronomer in the world. His telescope was 9.6 meters in diameter and had the now-familiar paraboloid shape; it could be steered to receive signals from various directions. Reber confirmed Jansky's observation of radio waves from the center of our galaxy, and he was the first to detect radio waves from the Andromeda galaxy and from the Sun. Between 1940 and 1948 he produced contour maps showing the intensities of radio waves in different parts of the sky. His work had an important influence on the later development of radio astronomy.

In 1942, during the Second World War, the British engineer James Stanley Hey (b. 1905) was investigating what was thought to be jamming of British radar by the Germans. He found that the maximum interference appeared to follow the movement of the Sun. An active sunspot was traversing the solar disk at the time, and Hey was able to show that it was responsible for the emission of radio waves of about one meter in length. In a sunspot region there is emission of electrons and other charged species moving at high speeds; when these pass through the powerful magnetic field of the Sun, radio waves are emitted. Hey later detected radar emissions from other galaxies and from meteors, and instigated procedures for tracking meteors, which could be done night or day. After the war there were many further advances in radio astronomy. Many radio interferometers were built on, essentially, the same principle involved in Thomas

Young's experiments (see page 79) on the interference of light. This equipment was highly effective in the study of radio emission from stars.

Radio waves pass freely through clouds, unlike radiation in or near the visible part of the spectrum. Since cloudy skies are only too common in Britain, optical astronomical spectroscopy was limited, and the initiative had largely been seized by the United States, particularly in observatories in the west and south. Radio astronomy, on the other hand, could be carried out in Britain as well as in the United States. Bernard Lovell (b. 1913) of the University of Manchester was particularly active in this area of research, tracking meteors and being the first to establish that their orbits are "closed," meaning that they are confined to the solar system rather than being of interstellar origin.

Early radio equipment was massive and could not easily be steered to receive radiation from a particular direction. Lovell organized the construction of a suitable paraboloid radio telescope at Jodrell Bank near Manchester. The dish was seventy-six meters (about 250 feet) across, and weighed about 1,500 tons. It was borne on turrets that had carried fifteen-inch guns on battleships, and could be steered to point in various directions. After many serious financial problems, this telescope was finally completed in 1957. The laboratory containing it is now known as the Nuffield Radio Astronomy Laboratory. It attracted early attention by tracking the carrier rocket of the world's first artificial satellite, the Russian-built *Sputnik*. The telescope has been used for tracking other satellites, and for studies of the Moon and the planets, but the emphasis has always been on the exploration of outer space by radio astronomy.

A number of other radio telescopes have subsequently been erected in various parts of the world. One of these is the James Clerk Maxwell Telescope, built on the top of a mountain at Mauna Kea, Hawaii, and opened in 1987 as a joint project supported by Canada, France, and Hawaii. Also at Mauna Kea is an optical telescope, the Canada-France-Hawaii Telescope.

Of course, valuable work continues with optical telescopes. One of these is the 200-inch (about five meters) telescope at Mount Palomar in California. Other optical telescopes are the Keck Telescope at Mauna Kea, with an optical diameter of 9.82 meters, and two 8.1-meter Gemini telescopes, one in Mauna Kea and the other in Cerro Pachon, Chile. Most of these large telescopes are supported jointly by a number of countries. One of the most impressive is the European Southern Observatory's Very Large Telescope (VLT), constructed in Cerro Paranal in the Chilean Andes by a consortium of European nations. It comprises four telescopes, each with an eight-meter mirror, which function together as a gigantic sixteen-meter telescope.

Also, optical and radio telescopes have been established in space. Valuable pioneering work was done by *Voyager 2* from 1977 to 1989, when it reached the planet Neptune. The famous Hubble Space Telescope was launched in 1990, but because of a series of technical difficulties became fully operative only in December 1993. It now makes observations of great importance in all fields of astronomy. The satellite *Hipparcos* is operated by the European Space Agency, and is primarily concerned with making precise measurements of the positions of stars and galaxies. Its name is partly a tribute to the great Greek astronomer Hipparchus of Rhodes (c. 170–c. 125 BCE), and partly an acronym of *hi*gh *p*recision *pa*rallax *col*lecting *s*atellite.

The discovery by radio astronomy of *quasars* was an advance of great significance. The word quasar is an acronym for *qua*si-*ste*llar *r*adio source. Quasars belong to a class of star-like celestial objects that exist at enormous distances from Earth. Some of them have also been identified by optical astronomy, and their important characteristic is a large red shift, emissions from the far ultraviolet being observed in the visible part of the spectrum; this shows that they are moving away from us at enormous speeds. The individual photons present in radio waves carry very little energy, but the quasars emit so many that the energy emitted is far greater than that emitted by any other cosmic source. The energy output is sometimes more than that of 10 billion suns, which means that they emit as much energy as an entire galaxy. This is attributed to their enormously high density. Quasars are believed to be an early stage of galaxies. They are related to *black holes*, objects consisting of extremely dense matter, which creates enormous gravitational fields. Anything coming close to a black hole is drawn into it, becoming part of the dense mass with the production of tremendous amounts of energy. The study of quasars gives us glimpses of events occurring in the very remote past.

Another interesting heavenly body is the *pulsar*. In 1967 Jocelyn Bell (b. 1943), a research student at Cambridge University and working under the direction of Anthony Hewish (b. 1924), noticed radio pulses emitted at regular intervals of about 1.3 seconds. Later in the year she and Hewish found other objects emitting similar periodic radiation, and today over five hundred of them are known. It was later realized that pulsars are neutron stars, whose existence had been postulated earlier. Theoretical studies suggest that the magnetic field of an assembly of neutrons the size of the earth is a thousand billion (10^{12}) times that of the earth, and that the radiation results from the rotation of this enormous magnetic field. The observed pulse is like a lighthouse: the star rotates rapidly and the pulse is observed corresponding to the passing of the peak of the emission. In 1974 Hewish received the Nobel Prize in physics for the discovery of pulsars; he shared the prize with Sir Martin Ryle, who also did pioneering work in radio astronomy. An interesting feature of pulsars is that although their emission is normally extremely regular, there is occasionally a temporary change of frequency. It has been suggested that a pulsar is like the earth, having a solid crust around a fluid core, so that occasionally there is slippage between the crust and the core.

Another important discovery made possible by radio techniques is the *cosmic microwave background*. Since this is particularly important in connection with the creation and age of the universe, we will consider it later.

The most recent work in astronomy has revealed that the size and complexity of the universe are vastly greater than previously imagined. There are estimated to be one hundred billion (10^{11}) galaxies, often disk-shaped with a typical diameter of 100,000 light-years. To get some idea of a number like 10^{11}, suppose that each one of them were given a designation consisting of about six letters or numbers. If you published a series of volumes listing no more than the designations, it would take three hundred sets of all the volumes of the *Encyclopaedia Britannica*.

There are many shapes and sizes of galaxies. Some are elliptical and much larger than our own, while some are much smaller. Our own galaxy has about 200 billion (2×10^{11}) stars in it, and has a mass equal to the mass of about 400 billion (4×10^{11}) suns; we say that its mass is 400 billion (4×10^{11}) solar masses. This is about average, so that the total number of stars in the universe is about 10^{21}. To illustrate this number, suppose we represented each star in the universe by an orange. If we piled 10^{21} of them together they would occupy a volume of about 10^{18} cubic meters (2.4×10^8 cubic miles). If these were piled on the United States, which has a surface area of about ten million (10^7) square kilometers (3.8 million square miles), the layer would be about 100 kilometers (62 miles) thick—higher than any mountain on earth.

Or, consider this: if our Sun were represented by an orange, the earth would be a pellet a millimeter (about three hundredths of an inch) in diameter, twenty kilometers (12.4 miles) away from the orange. On the same scale the nearest star would be 10,000 kilometers (6,200 miles) away, and the galaxy Andromeda, a member of our Local Group of galaxies, would be five billion kilometers (about three billion miles) away. The galaxies that are furthest away from us, about ten billion light-years away, would be 200 million million (2×10^{14}) kilometers (about 120 million million miles), or about twenty light-years, away from the orange representing the Sun. The galaxies contain unusually high concentrations of stars compared with the rest of space, and if every star in the universe were evenly dispersed over all space, the nearest star would (on the same reduced scale, with the Sun the size of an orange) no longer be 10,000 kilometers (about 6,200 miles) away, it would be many millions of kilometers or miles from the orange representing the Sun.

It is estimated that there are about 10^{78} atoms and about 10^{88} photons in the universe. It is believed that at the beginning of time there were no atoms, but a vast number of photons from which the atoms were formed. There are still about ten billion times as many photons as atoms in the universe. If we imagine splitting all the stars and planets in the universe into their constituent atoms, and spreading the atoms uniformly throughout the universe, there would be about one atom in every ten cubic meters of space. The diffuse gas between the galaxies also contains one atom per ten cubic meters. The total density of atoms in the universe is therefore one atom per five cubic meters, or 0.2 atoms per cubic meter. This figure is of great significance, since theory tells us that if the density of atoms were more than ten atoms per cubic meter, the gravitational attraction between them would eventually bring the expansion of the universe to a halt. Since the density is well below this critical density, it looks as if the universe will go on expanding for all time.

Astronomers have been able to estimate the amount of matter in the universe, and even say something about its nature. They make an estimate of the amount of matter from the extent of the various gravitational attractions. It is possible to deduce the mass of Earth and the Sun from the speed of planets orbiting around them, and, in the same way, the mass of galaxies can be deduced from the speed of stars orbiting within them. However, there is a difficulty, namely, that the mass of every galaxy appears to be larger than can be accounted for by the material now identified. There must be some additional, undetected substance present in the universe, without which some of the stars in the galaxies would fly away and the galaxies would fall apart. This additional material, amounting to some 90 percent of the mass of the universe, is called *dark matter*. Astronomers have no strong opinions about what it could be. One suggestion is *neutrinos*, which are particles that have been detected traveling through the universe in exceedingly large numbers, and which appear to have a tiny mass. The latest work on them, however, leads to the conclusion that they contribute only a fraction, perhaps a tenth, of the mass that seems to be missing.

Another suggestion is that dark matter is due to *brown dwarfs*, stars that are too small for nuclear reactions and that emit little radiation, easily escaping detection. Some astronomers claim to have discovered a brown dwarf present in our own solar system, moving in an orbit around our Sun but at an enormous distance from it. It may be that there are many brown dwarfs present in the universe, detectable only with great difficulty, but more investigation is required to discover the reason for the missing mass.

*   *   *

Nothing of much significance was done in geology until the final decade of the eighteenth century. It is true that a few interesting geological observations had been made earlier. Leonardo da Vinci (1452–1519), during his work on the construction of canals in Italy, examined the rocks and saw many marine fossils in them. He deduced that the mountains of northern

Italy had at one time been covered by the sea. Robert Hooke (1635–1703), who lived at about the same time as Isaac Newton, made studies of fossils and concluded, from the thickness of the beds in which they lay, that they must have been formed over enormous periods of time.

For many decades geology was a highly controversial science. Many people had strong preconceptions about the subject based on a literal interpretation of the Old Testament. Deductions from clear geological evidence were often rejected on the basis of these prejudices. A leading early geologist was Abraham Gottlob Werner (1749–1817), a professor at the mining academy at Freiburg. He was one of the first to realize that the earth's crust does not consist of a chaotic jumble of rocks but an ordered succession of layers, the oldest at the bottom and the youngest at the top. He gave support to the so-called *Neptunian* theory, according to which the mountains and valleys of Earth always had the same general form as they have today, but earlier were submerged in a thick ocean containing the substances in today's crust. These substances became deposited by a complex series of events, and the excess water eventually disappeared. There were several difficulties with this theory, one being that it didn't explain volcanoes. An alternative theory was presented by the so-called *Vulcanists*, who thought that at least some rocks had been formed at very high temperatures.

Important contributions were made by Scottish geologist James Hutton (1726–1797; see figure 19), who did much of his work in Edinburgh and numbered among his friends such distinguished men as chemist Joseph Black, philosopher David Hume, economist Adam Smith, engineer James Watt, and painter Henry Raeburn, who painted his portrait. Hutton made many geological observations of his own, mainly in Scotland, and formulated general principles that are now widely accepted. He realized that the biblical flood could not explain his geological observations, even if it had been of long duration. His theory, referred to as *Plutonism* or *uniformitarianism*, placed special emphasis on the role of rivers in excavating valleys and depositing dissolved and suspended material. He suggested that sediment carried by rivers would be washed into the sea and accumulate in deposits, which might then form new rocks by the action of heat coming from the interior of the earth. Perhaps under the influence of his friend James Watt, he regarded the earth as essentially a gigantic steam engine, whose subterranean heat created upheavals from time to time. Hutton denied the possibility that catastrophic events such as floods play any major role. He wrote of "a cyclic progression of changes so ancient as to obscure any vestige of a beginning and no prospect of an end." Hutton presented his work in a large book, first published in 1785 and expanded over the next decade.[6] At first there was much opposition to Hutton's theory, and for a time most geologists continued to accept the idea of *catastrophism*, believing that major catastrophes, such as the biblical flood, were necessary to explain the earth's development.

PHILOSOPHERS

Fig. 19. Scottish geologist James Hutton, *right*, in conversation with distinguished Scottish chemist Joseph Black (1728–1799). Often regarded as the founder of modern geology, Hutton established that the internal heat of the earth is important to rock formation. He was the first to abandon older ideas about geological changes, replacing them with his theory of uniformitarianism, according to which changes take place continuously over very long periods of time. (Contemporary cartoon. Courtesy of the Edgar Fahs Smith Collection, Annenberg Rare Book and Manuscript Library, University of Pennsylvania.)

The career of Oxford geologist William Buckland (1784–1856) is an interesting one, since the development of his ideas from catastrophism toward uniformitarianism parallels the change in geological thinking in his time. Buckland was an Anglican clergyman who later became dean of Westminster. As a churchman he initially favored catastrophism, following a *diluvial* theory that placed emphasis on the effects of the great flood of the Old Testament, and his early publications were written from this point of view. Buckland was one of the founding fathers of the branch of geology called *paleontology*, which deals with fossils. He made a particular study of the contents of Kirkdale Cave in Yorkshire, a large collection of teeth and bones that Buckland established as belonging to animals such as hyenas, elephants, rhinoceroses, and hippopotami. The general opinion was that they were evidence for Noah's deluge, the animals having taken refuge in the cave. However, Buckland produced convincing evidence that the other animals had been eaten by the hyenas. This removed any evidence for a flood, and Buckland's later studies made him more and more convinced that the diluvial theory was inadequate to explain the geological findings. Buckland was an intellectually honest scientist since, although in his earlier career he had strongly advocated the diluvial theory, he gracefully changed his mind on the basis of the evidence that he and others accumulated.

In 1840 Buckland became an enthusiastic supporter of the ideas of Swiss geologist Jean Louis Rodolphe Agassiz (1807–1873), who emphasized the role played by glaciers in forming the earth's surface. He concluded that much of northern Europe had at one time been ice-covered, and evidence has since shown that there were several ice ages. At a meeting in November 1840, Agassiz read a paper entitled "On glaciers, and the evidence of their having once existed in Scotland, Ireland and England." It was followed by a paper by Buckland, who acknowledged that he, too, had found compelling evidence matching Agassiz's, and he therefore withdrew his previous opposition to the theory. Others, however, remained hostile to Agassiz's theory, mainly for religious reasons.

Much support was given to both Hutton and Agassiz by Charles Lyell (1797–1875), who was born in the year of Hutton's death. He made an important contribution with his three-volume *Principles of Geology*, published from 1830 to 1833. It is interesting that this book was widely read, not only by scientists but by the educated public, had larger sales than most books by popular novelists, and ran to several editions. The book was consistent with Hutton's for the most part, but there were some significant differences in detail. For example, Hutton had thought that few changes in the earth were still taking place, but Lyell showed, by his direct observations, that the coast of Sweden was rising at the rate of several feet a century. It was thus possible for mountains to be formed gradually over long periods of time, rather than by catastrophic convulsions. After Agassiz gave his paper in 1840, and there was opposition to the glacial theory, Lyell asked: "If we do not allow the existence of glaciers, how shall we account for these appearances?" No one could give a satisfactory answer, and the glacial theory soon became accepted.

One particularly important geological theory that many scientists opposed at first was *continental drift*, according to which continents have changed their positions over long periods of time. Several geologists proposed this idea, but the first to build a convincing case for it was Alfred Lothar Wegener (1880–1930). He had obtained a doctorate from Heidelberg University in astronomy rather than geology, and was primarily a meteorologist. On the basis of several lines of evidence, including the distribution of fossils, he proposed that the continents may once have been joined up into one supercontinent, which he called Pangaea. He suggested that Pangaea broke up about 200 million years ago and the fragments drifted apart to form the continents as they are today. He first published this idea in a book in 1912; its English translation, *The Origin of Continents and Oceans*, appeared in 1924.

At first this theory was either ignored or opposed. The main objection to it was that it seemed impossible for the continents to move. Also, Wegener had been too specific in proposing a rate at which the continents spread, and when appropriate observations were made, his rate was shown to be wrong. As often happens with new theories in science, it was thrown aside by many because of incorrect details; in other words, the baby was thrown out with the bath water. A similar thing happened to Darwin with his theory of evolution; the few tiny points of detail on which he went wrong were taken by some people as evidence that the whole idea was wrong.

Further progress on this subject came partly as a result of the work of British geologist and geophysicist Arthur Holmes (1890–1965), a pioneer in the use of radioactive decay methods for dating rocks. In 1928 he suggested that convection currents within the earth's mantle might provide the driving mechanism for continental drift. Little attention was paid at first to his ideas, but in 1944 his influential book, *The Principles of Geology*, appeared and his arguments were convincing. Canadian geophysicist John Tuzo Wilson (1908–1993) was initially a staunch opponent of continental drift, but his later work supported it. One of his important contributions was the concept of the *transform fault*, which occurs when plates in the earth's crust slide past one another rather than one of them sinking below the other. This newer version of continental drift, called *plate tectonic theory*, deals with the formation, movement, and destruction of the earth's outer shell, which includes both the continental and oceanic crust. The lithosphere, formed by the solid crust of the earth, consists of large, tightly fitting plates that float on the semimolten layer of the mantle beneath them. The plates move relative to one another, and over the course of time this movement has brought the land masses into the present arrangement of continents and oceans. Continents drift at about the same rate as our fingernails grow.

All that we can discover directly about Earth relates to only a tiny fraction of it, the part that is no more than a few kilometers from the surface. The radius of Earth is about 6,400 kilometers (4,000 miles), but no one has ever been more than 4 km (2.5 miles) from its surface, and no one has bored deeper than 20 km (twelve miles). Information about the rest of Earth has to be obtained indirectly, and much of what we have learned has come from the study of seismic waves, waves set up by shocks. These can be natural, due to earthquakes, or artificial, from explosions. The shock waves are reflected and refracted by boundaries between different layers of rock, much as light is reflected and refracted from the surface of a block of glass. A number of seismic stations have been established throughout the world, and by analyzing seismic waves, it is possible to infer something about how Earth is constructed.

It has been established in this way that Earth has a solid inner core with a radius of about 1,500 km (900 miles), surrounded by a molten outer core with a thickness of about 1,900 km (1,200 miles), so that the total thickness of the core is about 3,400 km (2,100 miles). This core consists chiefly of iron, and its temperature is about 5,000 °C (8,600 °F). The reason that the inner core is solid and the outer core liquid is that the pressure is greater at greater depths, about three million atmospheres near the center of the earth. Above the core is the earth's lower mantle, considerably cooler, about 2,900 km (1,800 miles) in thickness, and solid in its deepest regions, again because of the high pressure. This solid region is also called the *mesosphere* (Greek *meso*, intermediate), a word that is also confusingly applied to one of the layers of Earth's atmosphere.

Above the mesosphere, and still a part of the mantle, is the *asthenosphere*, from the Greek *asthenes*, or weak. This is a region of weakness, where the rocks are partially melted and form a slush. Everything above the asthenosphere is called the *lithosphere*, which is about 300 km (200 miles) thick. Because of the slushiness of the asthenosphere, the material above it floats on it. This is the reason for continental drift. The upper part of the lithosphere is called the earth's crust, which in the land areas is up to 60 km (40 miles) thick, and is much thinner, 5 to 11 km (3 to 7 miles) thick, in the oceans.

*   *   *

As late as the twentieth century, biology was a very soft science. A vast amount of information had been collected and there were a number of well-established theories, but the subject still lacked the precision of physics and chemistry. This is partly due to the very nature of the subject. Biological systems are remarkable for showing variations from one another, because of their great complexity compared to the kinds of systems studied in any of the other sciences. In this chapter, since the biological information is so extensive, I will focus attention on the evidence relating to the evolution of species, and discuss how the theory of natural selection developed from the data. Since the middle of the twentieth century the subject has become transformed by the introduction of physical methods, such as x-ray determinations of the structures of some large and important biological molecules.

One of the first biologists, and some (like Darwin) would say the greatest, was Aristotle (384–332 BCE). He recorded a vast body of biological information, particularly relating to animals, and did a certain amount of classification (taxonomy). Over subsequent years a great deal of work was done on the classification of plants and animals. Living things can be divided into a number of groups, the members of each group having marked similarities to each other. Each group can be divided into sub-groups, which can be divided into species. The definition of a species is that its members can interbreed, while members of different species cannot. Today there are about thirty million species of plants and animals on Earth. There are about 10,000 different species of birds, and at least 300,000 species of beetles. British geneticist J. B. S. Haldane was once asked what the study of biology tells us about God. He replied, "He must be inordinately fond of beetles."

Religious writings such as the Old Testament suggest that every species on Earth was created separately and independently. The various and numerous species of animals and plants were assumed to be immutable, placed on the earth in their present form. John Milton, in *Paradise Lost,* gave a particularly specific and graphic account of the origin of life, and even today his influence is still detectable. According to Milton, the earth opened at the creation, and all land animals emerged, the lion pawing his way out of the ground, shaking his brindled mane. French naturalist George-Louis Leclerc, Comte de Buffon (1707–1788), however, suggested something quite different in his immense, twelve-volume work, *Histoire Naturelle.* As he collected information he noticed that there are marked similarities between members of different species, for example, between the dog, wolf, fox, and jackal. He drew the conclusion that they were members of one family and not created separately, especially since they showed similar anatomical features that seemed to serve no useful purpose. This obviously suggests a theory of evolution. A more explicit theory of evolution was put forward in the 1790s by Erasmus Darwin (1731–1802), Charles Darwin's grandfather. It had some unsatisfactory features, which Charles recognized.

An alternative theory of evolution was advanced by the Frenchman Jean Baptiste Lamarck (1744–1829), whose name now commonly refers to an error. Most of his work was done at the Natural History Museum in Paris. He arranged animals in a series, beginning with primitive organisms and continuing to the most complex. He suggested in about 1809 that there is a gradual process of development from the simple to the complex. He believed that changes brought about in the lifetime of an individual organism would be inherited by the next generation. In other words, he proposed the *inheritance of acquired characteristics,* which is now known as *Lamarckism.* His theory is simply explained by the example of giraffes. In order to survive, they eat the leaves of tall trees, and therefore stretch their necks. The next generation would, according to Lamarckism, inherit the longer necks, and from generation to generation the necks would get longer. We now know that this is wrong, and Lamark receives little credit for any of his ideas. This is unfortunate: he was a great pioneer, and his work was important in influencing others, including Darwin. There should be no shame in suggesting a theory that has to be overturned as a result of later work.

Lamarck's later life was an unhappy one. He suffered from poverty all his life and went blind, and his remains were thrown into a common grave. French naturalist Georges Cuvier (1769–1832) ridiculed Lamarck in a sarcastic eulogy, describing his ideas as worthless and unscientific. In fact, Lamarck's conclusion about the inheritance of acquired characteristics was not unreasonable, and was accepted by Darwin at first. Indeed, under some special circumstances, acquired characteristics can be passed on. For example, if a person is excessively exposed to x-rays, it can induce mutations that can be transmitted to future generations. Also, genes of a developing embryo can be affected by chemicals, such as drugs, in the surrounding fluid. Lamarck's idea was perfectly reasonable, but it is not the main factor in the evolution of species.

The theory of evolution that is now accepted was introduced later in the century by Charles Robert Darwin (1809–1882; see figure 20), whose conclusions were based on his observations during a five-year voyage as naturalist on the British naval vessel *Beagle*. On this voyage Darwin gained a detailed practical knowledge of the geology, plants, and animals of many distant lands. He shipped back to England a vast collection of geological and biological specimens. On his return home in 1836 he published some of the details of his observations, but was reluctant to publish his ideas about evolution, because he knew that they would produce a controversy. He had thought that Lamarck's idea of the inheritance of acquired characteristics was correct, but later decided that an important change was required. His alternative explanation for the long necks of giraffes was that if some giraffes had longer necks than others, the ones with the longer necks were more likely to survive, since they could reach more food, and so were more likely to have offspring. Those with shorter necks could reach less food and were less likely to survive long enough to produce offspring.

For over twenty years Darwin put off publishing his theory. His decision to do so resulted from his receipt in 1858 of a letter from Alfred Russel Wallace (1823–1913), a younger man who had also made a lengthy sea voyage. Darwin realized that Wallace had come to much the same conclusions about evolution, and that he was about to publish his theory, which would have won him priority. The two had some discussions about the matter and finally came to an amicable agreement. They arranged for separate papers to be read at a meeting in London on 1 July 1858 of the Linnean Society, an important scientific society dealing with biological matters of this kind. It is remarkable that neither of the two men were present on

Fig. 20. Charles Robert Darwin's theory of evolution by natural selection has had a profound influence on the progress of biology and on the lives of everyone today—including those who claim not to accept it! (Photograph taken about 1874.)

this historic occasion, and even more oddly, the announcement appears to have been largely ignored by other scientists.

In the following year, 1859, Darwin published *The Origin of Species by Means of Natural Selection*, which was at once widely read even by the general public, and continues to be read today. *The Descent of Man and Selection in Relation to Sex* appeared in 1871. His theory at once attracted attention and created considerable controversy among the public, although it was fairly quickly accepted by many scientists, who understood the strength of the evidence for it. Today Wallace's contribution to the theory has been somewhat forgotten, since Darwin's exposition of it was more complete and was supported by more evidence than that of Wallace.

Darwin's theory is often referred to as the *survival of the fittest*, an expression invented by the philosopher Herbert Spencer (1820–1903), a man of remarkable versatility who wrote on science, psychology, and philosophy. Darwin did not at first like the expression "survival of the fittest," preferring the term "natural selection." He later realized that the term natural selection was being misinterpreted as meaning that God controlled the process, the exact opposite of what Darwin meant.

The evidence for Darwin's theory is overwhelming, and among scientists there is disagreement only about minor details. The weakness of the theory in its original form is that it provided no mechanism by which evolution occurred. These began to be provided by experiments from 1856 to 1863 by the monk Gregor Mendel (1822–1884; see figure 21) in the gardens of a monastery in Brünn, then the capital of Moravia and now (called Brno) in the Czech Republic. Between 1856 and 1863 he grew about 30,000 pea plants, artificially fertilizing certain plants that had special characteristics. For example, he crossed tall plants with short plants, and showed that all the plants of the first generation were tall. In the next generation, however, some plants were tall and some short, in the ratio 3:1. He concluded that each plant receives one factor from each of its parents, tallness being dominant and shortness recessive. These factors were later called *genes*. Mendel's work remained unnoticed by other biologists for many years; Darwin had a copy of his publication in his library, but the pages remained uncut, so he apparently read little of the book. Mendel's conclusions have been supported by a great deal of independent work.

Over the years, evidence has accumulated that leads to a greater understanding of biological inheritance. In 1858, two years before Darwin's *Origin* was published, German pathologist Rudolf Virchow (1821–1902), working at the University of Berlin, showed that each living cell has a central concentration of material known as a *nucleus*. Soon afterwards scientists found that the nucleus of every cell contains structures called *chromosomes*, which always come in pairs. Before cell division, the pairs double and are then shared between daughter cells. The obvious conclusion, proved later, was that the chromosomes are the carriers of hereditary factors.

Experiments performed from about 1908 onwards by American biologist Thomas Hunt Morgan (1866–1945) were particularly significant. He carried out, first at Columbia University and then at the California Institute of Technology, an extensive series of investigations on *Drosophila*, tiny fruit flies that produce a new generation about every two weeks. Morgan confirmed and extended many of Mendel's findings, observing that certain pairs of genes tend to be inherited together more frequently than other

Fig. 21. Austrian priest and botanist Gregor Mendel discovered the basic statistical laws of heredity and provided a mechanism for Darwin's theory of evolution. He became abbot of his monastery in 1868, after which he had little time for research. (Courtesy of the Edgar Fahs Smith Collection, Annenberg Rare Book and Manuscript Library, University of Pennsylvania.)

pairs. He concluded that the genes are individual units, arranged in a particular order along a chromosome like beads on a string. He was thus able to achieve a certain amount of mapping of genes along a chromosome. Morgan was awarded the 1933 Nobel prize for physiology or medicine.

In the nineteenth century it had been concluded that genes sometimes undergo a change known as a *mutation*. American geneticist Hermann Joseph Müller (1890–1967) worked on the effect of x-rays on cells, showing that they induce genetic mutations and was awarded the 1946 Nobel Prize for physiology or medicine. He called the public's attention to the dangers of various kinds of high-energy radiation. Previously, x-rays had been used much too freely in hospitals, without proper precautions, and x-ray equipment was even used in shoe stores to test whether a shoe fit a customer!

In 1944, Canadian-born Oswald Theodore Avery (1877–1955) and his colleagues at the Rockefeller Institute in New York were working on *pneumococci*, in particular on the transformation of a nonvirulent form into a virulent form, and found that the transformation is brought about by a substance called deoxyribonucleic acid (DNA). This suggested that the genes were portions of DNA molecules, which was subsequently found to be true of the genes of every living organism, plant or animal. Crick and Watson later determined a helical structure for DNA.

Important work was being done at about the same time on *bacteriophages*, which are viruses that infect bacterial cells. The German-born biophysicist Max Delbrück (1906–1981) worked in this field, particularly at the California Institute of Technology. He had been educated as a nuclear physicist and had studied in Copenhagen with Niels Bohr. Another who worked on bacteriophages was Alfred Day Hershey (b. 1908) of the Carnegie Institute in Washington. At the University of Indiana, Italian-born Salvator Edward Luria (1912–1991) obtained good electron micrographs of bacteriophages. All three obtained results that supported the conclusion that genes are portions of DNA molecules, and they shared the 1969 Nobel Prize for physiology or medicine.

*   *   *

By the 1950s scientists were convinced that genes played a crucial role in the evolutionary process, and that they were groups of DNA molecules. They knew that a single gene carried a vast amount of genetic information, and they strongly suspected that this was determined by the order in which four chemical groups, known as *bases*, were arranged along the gene's molecular chain. So it was important to determine the structure of a DNA molecule, in particular the arrangements of the four bases. By this time, the technique of x-ray spectroscopy revealed the structures of some large biological molecules. During World War II, Dorothy Crowfoot Hodgkin (1910–1994) found, by x-ray technique, the exact structure of penicillin. This was before it had been done by conventional chemical methods. Several scientists determined the structures of individual protein molecules, and in 1948 the American chemist Linus Pauling (1901–1994) made the suggestion that helical (spiral) structures sometimes occur in protein molecules.

Before the work on the structure of DNA, determining the structure of a large molecule took painstaking experimental x-ray work, followed by detailed mathematical analysis of the results—work that is now done much more easily by computers. How the structure of DNA was arrived at is quite different, making less use of the x-ray data. Instead, a plausible molecular model was suggested, partly on the basis of experimental results, but also considering how the various atomic groups fit together. An important part of the work was building models of molecules.

Rosalind Elsie Franklin (1920–1958), at King's College, London, did a great deal of excellent and painstaking work on DNA, and her results were crucial to the final solution of the problem. Also doing excellent x-ray work on DNA at King's College was Maurice Hugh Frederick Wilkins (b. 1916), who was born in New Zealand. Another who was interested in the DNA problem was Francis Harry Compton Crick (b. 1916), who worked in the Cavendish laboratory at Cambridge. All three had received most of their education and experience in physics and in the more physical aspects of chemistry. The fourth person involved in the DNA work was American James Dewey Watson (b. 1928), who was primarily a biologist and then working at the Cavendish laboratory. At the time the director of that laboratory was Sir William Lawrence Bragg (1890–1971), who had shared the 1915 Nobel prize for physics with his father Sir William Bragg (1862–1942) for their pioneering work on x-ray crystallography. At first, Bragg did not encourage the speculations of Crick and Watson on the structure of DNA, especially since Crick, who had not yet obtained his Ph.D. degree, was neglecting the work he was supposed to be doing. Later, however, when the DNA work promised to lead to a structure, Bragg encouraged them to proceed.

Since Pauling had shown convincingly that parts of proteins are in the form of a helix or spiral, Crick and Watson thought that DNA was likely to have the same kind of structure. They had also seen some of Rosalind Franklin's x-ray photographs, though without her knowledge, and these suggested a helical structure. Erwin Chargaff (b. 1905) at Columbia University had shown that there was a striking relationship between the amounts of the four bases that occur in DNA. The bases are adenine (A), guanine (G), thymine (T), and cytosine (C), and the number of units of A was always equal to the units of T, while the number of units of G was the same as the units of C. This is referred to as Chargaff's rule.

Crick and Watson devoted much effort to building molecular models, just as Pauling had done to arrive at his helical structure for proteins. The four bases have different sizes, and they had to make sure that everything fit together satisfactorily. In 1953 they proposed a DNA structure that involves a double helix, two helices intertwined. The structure can be visualized as a double spiral staircase, one reserved for walking upstairs, the other for walking down.

From models, Crick and Watson saw that the four bases were of different sizes, and that for the double helix to be of uniform width, an A unit could only be associated with a T, and a G must pair with a C. This pairing explained Chargaff's rule, and only in that particular pairing would the helix be uniform in width; any other pairing would give bulges along the chain, which was excluded by the x-ray work.

Replication involves the unwinding of the helix. Each single helix serves as a template for the formation of another helix. Wherever there was an A, a T would be become attached to it, and vice versa; a G always became bound to a C, and a C to a G. Helix-1 from the original double helix would thus bring about the formation of a new helix-2, while the original helix-2 would form a new helix-1. The two formed the double helix again, and replication was complete.

The announcement in 1953 of the double helix structure aroused great interest and excitement, and the work has been called the most original of the twentieth century. Although the structure had not been based heavily on experimental evidence, it seemed to be so plausible that it could hardly be wrong. As further experimental evidence accumulated, the Watson-Crick model became universally accepted. Crick, Watson, and Wilkins shared a Nobel Prize in 1962. Rosalind Franklin, whose experimental work had been so important in leading to the structure, had died of cancer four years earlier, and Nobel prizes are never awarded posthumously. Many people have felt that her work was not properly recognized when the prizes were awarded, and the fact that her experimental results were seen without her consent or knowledge was certainly unfortunate. Happily, she received some recognition later. Her Cambridge College, Newnham, dedicated a residence in her name, and in 2000, King's College, London, attached her name, together with that of Wilkins, to a new molecular biology building.

*   *   *

The field of *molecular genetics*, which is the application of the methods of physics and chemistry to genetics, progressed more rapidly after Watson and Crick proposed their helical structure for DNA. Much effort has gone into understanding the chromosomes, which occur in animal and plant cells, transmit hereditary information, and consist of strings of genes, which are segments of DNA molecules. A gene is a specific section of a DNA molecule that corresponds to a recognizable unit that plays a role in inheritance and metabolism.

Genes have many remarkable features. They perform two important functions, both crucial to life. In the first place, they are involved with the *replication* (copying) of an organism. Second, they control the organism's day-to-day survival. Once an organism is born the cells proliferate, and each one contains a copy of the individual's set of genes, known as the *genome*. There is nothing intrinsic to a specific gene, in the sense that when a gene is outside the cell there is no distinction between a bacterial gene, a plant gene, and a human gene. We are not able to say, on examining an isolated gene, what species it came from. Variants of particular genes are known as *alleles*. It is commonly said, even by experts in the field, that there are "genetic" differences between different people, but this is not strictly correct; all humans of the same sex have the same genes, and it is better to say that there are *allelic* differences.

We now know that it is the protein molecules rather than the chromosomes that are concerned with the day-by-day functioning of a living system. A protein molecule is a long chain molecule with a chain of amino-acid molecules. There are twenty different amino acids and each individual protein has a particular sequence of amino acids. Many biological structures, such as muscle and skin, are made up primarily of proteins, and it is mainly the proteins that are concerned with physiological activity. In the human body, for example, the protein hemoglobin transports oxygen from the lungs to the places where it is needed. The enzymes, which are proteins, are the biological catalysts; the stomach and the small intestine contain various enzymes that bring about the digestive processes. Most of the chemical processes that occur in living systems would occur too slowly if the appropriate enzyme were not present.

One fundamental problem that molecular geneticists had to attack was how the genes were able to control the manufacture of proteins. The mystery began to be solved in the 1960s. The order of the bases in DNA is the critical feature. Suppose that in a particular DNA molecule the order of the four bases T, C, A, and G happens to be

TTCGGTCGC . . .

One suggestion put forward at first was to consider the bases in pairs; to indicate this we can put commas between the pairs:

TT, CG, GT, CG . . .

It is easy to see that this cannot work; there are twenty different amino acids, and if we take four bases in pairs we can only get $4 \times 4 = 16$ possibilities:

TT, TC, TA, TG
CT, CC, CA, CG
AT, AC, AA, AG
GT, GC, TA, TG

Suppose that we consider triplets instead, so that, with commas, the sequence becomes

TTC, GGT, CGC . . .

We can now get $4 \times 4 \times 4 = 64$ possibilities, which is more than enough. This turned out to be the right answer; often different triplets lead to the same amino acid.

South African–born Sydney Brenner (b. 1927) made important contributions, particularly to understanding how the genes work. For a time he worked closely with Crick at Cambridge, and introduced the word *codon* to refer to a triplet of bases, such as TTC, in DNA. The essential feature of the coding is that there is an exact correspondence between the sequence of the codons in the DNA of the gene and the sequence of amino acids in the protein produced. For example, the codon TTC specifies the amino acid called phenylalanine, GGT specifies glycine, and CGC specifies arginine. The sequence TTC, GGT, CGC thus leads to the synthesis of a protein having the sequence: phenylalanine-glycine-arginine. All twenty amino acids could be coded without using all of the combinations, but in fact, several different codons lead to the same amino acid. Glycine, the simplest of the amino acids, is given by any of the following four codons: GGT, GGC, GGA, and GGG. The function of certain codons is to make the synthesis start or stop.

Besides Crick and Brenner, many others were involved in breaking the genetic code—that is, in relating the codons to particular amino acids. Severo Ochoa (b. 1905) of New York University and Arthur Kornberg (b. 1918) of Stanford University discovered the mechanisms of synthesis of DNA and ribonucleic acid (RNA), and were awarded the 1959 Nobel Prize for physiology or medicine. Marshall Warren Nirenberg (b. 1927) of the National Institutes of Health in Bethesda, Maryland, was the first to identify a codon with a particular amino acid. Har Gobind Khorana (b. 1922) of the University of Wisconsin and Robert William Holley (1922–1994) of Cornell University also took some of the final steps in unraveling the genetic code, and the three of them shared the 1968 Nobel Prize for physiology or medicine. Sydney Brenner rather belatedly received a Nobel prize for physiology or medicine in 2002.

An interesting project began in 1963 and continued successfully for many years. Brenner suggested that it would be profitable to focus attention on a tiny nematode (roundworm), which is composed of only 959 cells. This became known as the Worm Project, and by 1986 the main genetic and physiological details of the organism had been worked out. The Human Genome Project, of much greater complexity, came a little later.

A few other details about genetic coding deserve brief mention. The DNA in a cell nucleus does not form proteins directly, but remains in position and makes a copy of itself. The information is actually transmitted through an intermediary called messenger-RNA (mRNA). Ribonucleic acid (RNA) molecules, of which mRNA is a special form, are important constituents of cells. They are formed from DNA so that the bases are arranged in an equivalent order to that in the DNA molecule, but some details of their structure are different.

X-ray studies of protein molecules have always shown that a particular protein normally exists in a unique three-dimensional form; the one-dimensional strands are always folded in the same way. In a globular protein the long molecules are folded up like a ball of wool, but always in exactly the same pattern. This raises the question of how a molecular template could directly produce such a complicated three-dimensional structure, always in a unique pattern. The answer is that it does not do so directly. The protein molecule may be synthesized as a long strand of amino acids, but when it is surrounded by water molecules the molecular strand folds itself spontaneously into a particular shape. The precise sequence of amino acids determines the folding that takes place. The biological synthesis of a protein thus creates the molecule as a long strand of amino acids. Sometimes, as it is being formed, it steadily folds into its unique three-dimensional shape. In other cases, it folds into its final shape at a later stage.

There is one rather remarkable feature of the folding processes. Each protein has a specific conformation when it folds itself naturally, and if a protein is deliberately unfolded it will spring back into its specific conformation in a matter of a second. You would expect to be able to reproduce these folding processes on a computer, but this has not been achieved for any protein. Obviously, a delicate balance of forces is involved in the folding processes, and we still do not understand the details. The same must be true of the folding of chromosomes so they fit into cells.

A stupendous amount of information can be stored in a single cell. A cell in a human body, for example, can store about ten times as much information as in all thirty volumes of the *Encyclopaedia Britannica*. A bacterial cell has a much smaller capacity, by a factor of about 1,000, and could only store the information contained in the New Testament. In his fascinating book, *The Blind Watchmaker*, Richard Dawkins has suggested a prolific way to replicate the New Testament.[7] It could be encoded into a bacterium, which would reproduce and form ten million copies a day. He points out a few serious snags, however. Encoding the bacterium might take many years, and bacteria are extremely hard to read.

The nucleus of a typical human cell contains a string of DNA units distributed over twenty-three pairs of chromosomes. The length of the DNA from one of these chromosomes is about five centimeters (two inches). This means that the entire length of the forty-six chromosomes, if laid end to end, is about two meters (six feet). The neat packing required to get twenty-three pairs of chromosomes into the nucleus is best appreciated by an analogy.[8] Suppose we magnify the nucleus to the size of an average suitcase, about a million times. A single chromosome would then have a length of fifty kilometers (about thirty miles) and a thickness of about one millimeter (about three hundredths of an inch). Imagine packing a suitcase with (among other things) forty-six pieces of string a millimeter thick and fifty kilometers long! The cell does this so that each one of the codons in the forty-six filaments is accessible and able to carry out its functions in the cell. This is surely one of the greatest wonders of nature.

Biologists are fond of pointing out that if all the DNA in all the cells in a single human being were stretched out it would reach to the Moon and back eight thousand times (or to the Sun and back 250 times). This is true, but with all respect, hardly remarkable. If all the molecules in a liter of water formed a filament it would reach to the Moon and back twelve million times. These enormous lengths have nothing to do with the lengths of chains; they follow from the thinness of filaments. What *is* really remarkable in the tiny human cell is that a *single* chromosome is about five centimeters (two inches) long and that the entire length of the forty-six chromosomes is about two meters (six feet). Even more remarkable is the fact that they can be packed into the tiny nucleus of a cell, and that, when packed, they can still carry out so much chemistry.

All human beings of the same sex, regardless of race, have the same genes, of which there are about forty thousand. We only differ from one another as a result of minor variations in a fraction of a percent of our genes. The Human Genome Project, supported internationally, explores some of the details of human genetics. Each human genome contains about three billion base units, and it is hard to appreciate the enormity of this number. Imagine magnifying the genome so that it is 100 kilometers (about sixty-two miles) long. On that scale, *each centimeter* would contain about 300 base units (or, if you prefer, each inch would contain about 750 of them).

Only a small proportion of the DNA chains are functional, in the sense of encoding protein production. About 95 percent of the base pairs perform no useful function at all. They are simply errors of copying, or result from the evolutionary process. We again may be helped by thinking in terms of the 100-km (sixty-two-mile) road that represents the genome. If we traveled along it we would find the trip quite tedious. About a third of the road is covered with sequences that are just useless repetitions of previous sequences. About 5 percent of the road consists of countless repeats of blocks of five or more units. Although these strange features appear to have no functional significance, they cause each individual to be unique, and make it possible to identify people by genetic fingerprinting.

The initial objective of the Human Genome Project was to identify all of the three billion or so base pairs that exist in a given set of chromosomes. This part of the project was completed early in the year 2000. This list of base pairs belongs to an average person; individuals show small variations. A printout of the list would run to about three hundred very heavy volumes. To most of us, such books would be the most boring ever written, but experts find the genetic map of absorbing value and interest. A number of genes have been identified, but much work remains to be done. The gene that causes about half of us to be males, for example, starts with the following codons:

GAT, AGA, GTG, AAG, CGA, . . .

There are 240 letters in the complete gene, which is situated on the Y chromosome. Humans who lack this Y chromosome and, therefore, this male gene, are females.

There has been criticism of the Human Genome Project, particularly because it has proved very expensive and has siphoned money that might have been used for other kinds of research. A common argument takes the form of an analogy. Suppose we were studying the way in which a particular country worked, and began by making a painstaking and accurate inventory of every brick and stone in that country. Everyone would agree this is a complete waste of time. Only a superficial survey of the buildings in the country would be necessary, and we could devote our efforts to more significant matters. The analogy, however, is a bad one, since the positions of individual bricks and stones are not crucial to how the country runs; we could remove and exchange stones and even a few buildings without any significant effect on the country's economy or politics. With the genome, on the other hand, it is an entirely different matter. A trifling change in a few base units can make all the difference between a person in good health and one afflicted with a crippling disease such as cystic fibrosis. To gain a full understanding of the human genome will take much more time. We need to know how the individual genes are related to the details of the genome, and the relationship between the genes and the human condition. A great deal more remains to be investigated.

One fact that complicates the understanding of genetic problems is that the human genome, with its three billion units, does not correspond in any simple way with the picture that has emerged from studies of family relationships. There obviously is a relationship between the genes and the genome, and between the genes and the human condition, but it is complex.

The way characteristics are passed from one generation to the next is now beginning to be understood. It would make things easier if the traits handed down from generation to generation were controlled by single genes, but this is rarely the case. Single genes are found to control a few diseases, such as cystic fibrosis and muscular dystrophy, and in some cases the genes responsible have been identified. It may eventually be possible to develop techniques for repairing damaged genes.

More often, there is not a simple one-to-one relationship between a gene and a physical characteristic or a genetic disorder. An example is eye color. It used to be assumed that this would be controlled by a single gene, but in fact, four to six genes seem to be involved. The same is true of skin color. Most characteristics in humans are controlled by several genes, and gene interactions are often complex. Mendel's experiments with peas involved a much simpler form of inheritance than is usually found with higher animals.

When many genes play a part in a genetic disorder, controlling it may be much more difficult. This is the case with heart disease, cancer, and diabetes. With cancer, over 100 genes have already been implicated, and there are probably more. To halt the onset of such disorders before they start is a task of great technical difficulty.

The science of molecular genetics has helped in the matter of race. For centuries, the popular opinion was that that there exist races of people, usually defined by their skin color and other physical characteristics, who at some prehistoric time had quite different origins. The human race was divided into biologically pure lineages that originally were distinct from one another, but which over the centuries have become somewhat blended by interbreeding. Religious respectability has sometimes been given to such ideas, for example, in the Judaeo-Christian tradition, by the suggestion that the black, white, and yellow races are descended from Ham, Shem, and Japhet, the three sons of Noah. There has always been a tendency for people of one race to despise those of another race, and this has had a disastrous social impact. In many parts of the world, and in the United States until a few decades ago, strictly enforced laws kept the white and black "races" apart.

Racist attitudes were brought out clearly with the hereditary condition known as Down's syndrome, in which, because of a chromosomal error, the sufferers are of diminished intelligence and have an unusual facial appearance. English physician Langdon Down (1828–1896) discovered the condition in 1866 and called it Mongolism, believing that the individuals had somehow slipped down the evolutionary ladder to resemble the Mongols of central Asia, who were considered a lower form of life. In Japan, Down's syndrome was denoted by a word that means "Englishism." Modern English dictionaries say that the word Mongolism is offensive; perhaps modern Japanese dictionaries make a corresponding comment about Englishism.

The idea that there is a hierarchy of human beings, with white people of high pedigree at the top, is not supported by modern genetic research.

The visible variations between different groups of people, such as color of skin, are due to differences in a small group of genes, corresponding to only a tiny fraction of the total number. The genes determining color do not appear to be linked to a significant number of other genes, including those relating to intelligence. There is thus no reason at all to believe, for example, that people of one race are more or less intelligent than those of another.

If we compare a group of white people from different countries to a similar group of black people, we find that, statistically, the genetic differences between a randomly selected white person and a black person are only slightly greater than those between two white people or two black people. The genetic differences between two white people from two European countries will be not much less—perhaps only 5 or 10 percent—than those between a white person and a black person. The differences between the majority of people in a country and the visible minorities thus lie largely in visible characteristics. We are all cousins under the skin.

The identity of a nation depends only on a shared history during a period of time that is insignificant, after all, compared with the period that *Homo sapiens* has existed. Cultural differences account for most of the differences between races.

*   *   *

Mitochondria, tiny, lozenge-shaped structures called organelles, are present in the thousands in each of the cells of our bodies. They are enormously important in our metabolism, since they bring about many chemical reactions that are essential to life. Mitochondria in humans are transmitted by the mother only; hardly any mitochondria from male sperms reach the egg at fertilization. A practical consequence is that it is possible to make identifications of persons through the female line. Genetic tests of this kind established the identity of the remains of members of the family of Nicholas II, the last czar of Russia. Because mitochondrial genes evolve more rapidly than other genes, they are often used in the study of the evolutionary process.

In a similar way, it has been established that the latest common ancestor of all humans along the female line was a woman—sometimes called Mitochondrial Eve—who lived a few hundred thousand years ago. On the basis of human remains found in Africa, it seems likely that Mitochondrial Eve lived on that continent.

*   *   *

As a direct result of work in molecular genetics, an important new field, genetic engineering, has emerged. This continuously expanding field has many applications and advantages, and inevitably some drawbacks. It is important to farmers, since much disease on farms can now be controlled. Artificially produced viral genes are being used to inoculate crops and they limit the ability of invading viruses to replicate within the plant cells. Also, when certain fungal-resisting genes are injected into clusters of corn cells, they stimulate them to replicate and grow much more satisfactorily than the original corn. There are many more examples of useful techniques of this kind.

Genetic engineering is also extensively used in medicine. Many drugs are difficult and expensive to extract from natural substances or to produce by ordinary chemical means, and genetic techniques may be of great value in their production. Such procedures have three important advantages. The procedures are usually simpler and cheaper to carry out. Second, drugs produced in this way are usually purer than those produced chemically or extracted from biological material. Third, human genes can be used in the processes, so the drugs are more compatible with humans. An important example is the production of insulin, which is widely used to treat diabetes; animal insulin sometimes causes allergic reactions, but insulin produced using human genes does not. Penicillin is another example. The original strain of *Penicillium notatum* used in the 1930s yielded only minute traces of penicillin from enormous vats of fungus. With much painstaking work over a number of years, biotechnicians altered the genetic character of the bacterium to produce larger yields of the drug. The work involved procedures, like causing mutations by radiation, that have sometimes horrified people. The results were highly successful, however, and necessary so that people could be treated with penicillin.

The drug interferon presents a particularly significant example. In 1957 scientists found that interferon is produced by cells in the human body in response to the attack of viruses. The substance produces a protein that stimulates the immune system, thus inhibiting the spread of infection. Its importance was recognized at once, but interferon could not be marketed because of its scarcity and high cost. Many thousand human donors were required to provide enough of the drug for a single patient, and a single dose would have cost perhaps $50,000. With the genetic engineering of the 1970s, the situation changed dramatically. In 1980 a gene for human interferon was introduced into bacteria, the first time this had been done for a human gene. By cloning vast numbers of bacterial cells from a small sample, it was possible to produce a cheap supply of the drug, which is now widely used not only to combat viral infection, including the common cold, but as an anticancer drug.

An important point about all this is that many people think "natural" food and drugs are more reliable than those produced "artificially." However, with modern techniques such as bioengineering, the truth is usually the other way around.

Recently there has been much progress in the treatment of genetic diseases like cystic fibrosis and muscular dystrophy by gene replacement. A single gene is responsible for each of these two diseases and it has been located. Various techniques are being used to insert a corrected version of a gene into a person's body. There has been criticism of such techniques, which some think unduly interfere with nature. Most of us would agree that creating "designer babies" by genetic techniques is undesirable, but it is difficult to see any moral objection to the treatment of genetic disorders. It would seem to me that, on the contrary, it would be immoral to fail to cure such diseases if it is technically possible to do so.

*   *   *

The theory of evolution is now accepted by all scientists, who regard its validity as being as sound as that of other great scientific theories, such as the reality of atoms and molecules, the quantum theory, and the theory of relativity. It is not, however, so universally accepted by the public, because some people think that it conflicts with their religious beliefs. Such a conflict is unnecessary. The evidence for evolution is so strong that the theory is as close to the truth as we can ever be in science. It is useful to summarize the main arguments for the theory.

Scientists do not claim to reach any absolute truth; we are merely concerned with what *works*. A theory is accepted when it explains all the known facts, obtained from observation and experiment, and equally important if it can be used successfully to predict what will happen under another set of conditions. Darwin himself arrived at his theory of evolution because he knew a great many geological and biological facts, observed during his five years on HMS *Beagle*. During subsequent years, a vast number of new relevant facts have been discovered. They have required a few modifications of Darwin's original theory, as almost always happens when a new theory is proposed, but everyone who is familiar with the subject is satisfied that the theory must be correct in its general outline.

An even more compelling reason for believing the truth of the theory is that the hard science on DNA and other genetic materials has all been completely consistent with evolution theory. The information that tells a cell how to function is coded in certain specific chemical units along a DNA molecule. The order in which these units occur has been determined, as has been the genetic code, by which the information in the genes is used in the synthesis of proteins. What is remarkable is that the same genetic code applies to all bacteria, plants, and animals that have been studied. This strongly indicates that all of them have a common origin; the probability that this identity could have occurred by chance is exceedingly small. (This theme is taken up in a little more detail in the next chapter.)

Also, long stretches of DNA have no function, and are clearly the relics of the evolutionary process.

Another strong argument in favor of evolution is that there are close resemblances between the arrangement of the genes in species that are closely related. Humans and chimpanzees, for example, are closely related, and their genetic makeup differs in only minor details; 98 percent of the genes of a chimpanzee are identical to those of a human. We have forty-six chromosomes, while chimpanzees have forty-eight, and the sequences of codons in the chromosomes are quite similar. Distantly related species, like humans and mosquitoes (who only have six chromosomes), show greater differences in the sequences of codons. All known genetic information of this kind is entirely consistent with Darwin's basic theory.

Finally, biologists who work with bacteria and viruses find that they evolve all the time. The influenza virus undergoes mutations to resist antibiotics. Here is evolution occurring in front of our noses.

It is a curious irony that so many people today who express disbelief in the theory of evolution greatly profit from the practical consequences of that theory in the food they eat and the medical treatments they receive. If the nonbelievers in the theory were all deprived for a short time of the practical consequences of the theory, I am sure there would be a massive shift of opinion to the side of evolution.

Chapter 5
How It All Began

> It does not at present look as though Nature had designed the universe primarily for life; the normal star and the normal nebula have nothing to do with life except making it impossible. Life is the end of a chain of by-products; it seems to be the accident, and life-destroying radiation the essential.

> —Sir James Jeans, *The Wider Aspects of Cosmogony*, 1928

Some of us can lead perfectly happy and contented lives without thinking much about how the universe came to be created, how our Earth was formed, how life began on Earth, or how our particular species of *Homo sapiens* came to exist. Some of us, however, cannot help wondering about such things, even though we realize that we can never (at any rate, in our lifetimes) reach a completely satisfying explanation.

Some religious people, and even some who no longer practice their religion, accept literally the explanations given in religious writings. The book of Genesis, the first book of the Old Testament, is taken directly from the Jewish scriptures. It gives a detailed account of the creation of the universe and of the creation of animals. It then gives two incompatible accounts of the creation of Adam and Eve, the first human beings; in one of them they appear together; in the other, Eve was formed from Adam's rib. Some Jews and Christians take the Old Testament as divinely inspired, to be taken literally.

Obviously there are serious discrepancies between these scriptural accounts and the evidence of three sciences: astronomy, geology, and biology.[1] Although some scientists adhere to religious beliefs, very few reputable scientists take a literal view of scriptural writings about the creation of the earth and of human beings. Religious people who have made an unbiased assessment of the scientific evidence also regard the scriptural versions as merely representing the thinking of the time. After all, it would have been inappropriate for God to have inspired the earthly writers of many centuries ago to give an account that was compatible with modern scientific conclusions; such an account would have been completely incomprehensible even to the most learned scholars.

Many religious leaders and theologians take this point of view. Even the Roman Catholic Church, never hasty to admit new ideas, has officially accepted many modern scientific ideas—with the reasonable proviso that they must be based on reliable evidence. In 1996 Pope John Paul II issued a statement accepting that "masses of evidence render the application of evolution to man and the other primates beyond serious dispute."

Before looking at some of the astronomical, geological, and biological evidence that supports modern scientific conclusions about our past, it is helpful to refer to figure 22. There is now compelling evidence, from a variety of sources, that the age of the universe is at least twelve billion years. It is hard for us to appreciate such an enormous period of time. In his excellent and fascinating book, *Before the Beginning*, Sir Martin Rees writes that a person who wanted to walk across the United States with the idea of arriving at the other side in twelve billion years would have to take *one step every two thousand years!*[2] Suppose some lethargic man walked from New York to Los Angeles at that pace, stepping back in time, so that as he trudges along he sees what is happening in earlier times. After only a few steps he has reached prehistory. Along the first few kilometers he sees a few beings who look human, but well before he has gone 100 kilometers (about sixty-two miles), there are no more. Later, he sees various odd-looking animals and plants, but by the time he is one quarter of the way across the country he observes only primitive living cells, and after a few more steps there are no more of those. After about one third of his trek he notices that our Earth is cooling and that rocks are forming. While still in Nevada, he sees stars formed in our galaxy. During the rest of the trip he is surrounded by a whirling cloud of hot gases, and at the end he gratefully collapses into a dense mass at a temperature of many billions of degrees.

Astronomers necessarily deal with very distant objects, so far away that we can never hope to make direct contact with them. In the case of the Moon and a few planets, it is possible to bring samples from their surface back to Earth, where experiments can be done on them. Most of the heavenly bodies, however, are completely out of our reach, and information about them can be obtained only from a great distance. The science is

Fig. 22. A trek across the United States, giving us some idea of the times that events occurred in our universe. The traveler takes one step every two thousand years, reaching the destination in about twelve billion years. The numbers show the times, in billions of years. Note that the time that human beings (*Homo sapiens*) have existed, about 100,000 years, is negligible compared with the age of the universe, at least twelve billion years.

therefore largely observational rather than experimental. However, the theories that astronomers propose on the basis of their observations can, in some cases, be tested by laboratory experiments, such as experiments on the nuclear reactions that are believed to occur. The circumstantial evidence for one particular theory of the creation of the universe, the big bang theory, is particularly strong.

In the last chapter we began to consider some of the important contributions of distinguished American astronomer Edwin Powell Hubble (see figure 18 on page 106), who observed and studied many galaxies far outside our galaxy. He discovered red shifts in the spectral lines of these galaxies. Red shifts are examples of the Doppler effect; they show that the source of the radiation is receding from us. From the red shifts, Hubble was able to obtain accurate estimates of the speeds of recession. He discovered a simple relationship between the speed of recession of a galaxy and its distance from Earth. The relationship is *proportional*, so if, for example, a galaxy is twice as far from us as another one, it is receding twice as fast. The speed of recession of a galaxy is thus related to its distance from us by the formula:

Speed of recession = Hubble's constant × distance

The main difficulty in obtaining the value of the Hubble constant is that, whereas the rates of recession can be obtained accurately from the Doppler shift, the distances are more difficult to estimate accurately, especially for the distant galaxies. Recently, quite reliable values have been obtained by satellites.[3]

The galaxies are gradually slowing down because of the gravitational attractions between them, but the effect does not greatly affect our estimates of the age of the universe. Suppose we go back in time, which means that we imagine that the speeds of the galaxies are all reversed. The ones that are far away were moving away from us at the highest speeds, so now they are coming back at high speeds. The nearer ones are coming back at lower speeds, and eventually they will all be at the same place. In other words, at some time in the far distant past, they apparently were all in one place. The obvious conclusion is that the universe was created by an enormous explosion that, as in all explosions, propelled some fragments more rapidly than others. Those propelled at the highest speeds have traveled the longest distances and are at the outer extremities of the universe. From the size of the Hubble constant, it is easy to calculate that the explosion must have occurred at least twelve billion years ago (see note 3). From now on we will use this conservative figure, though the true age is probably greater.

Therefore, we are led to the conclusion that the universe is expanding. It began with highly dense matter that exploded twelve billion or more years ago, and the fragments traveled in all directions at immense speeds. This is not to suggest that Earth is at the center of the universe, and that everything else is rushing away from us. If the universe was a lump of dough with currants in it, as we placed it in an oven it is cooked, the dough swells, and the distance between every pair of currants increases. In the same way, as our universe expands, the distance between almost every pair of galaxies increases. There are a few exceptions. For example, our neighboring galaxy, Andromeda, is approaching us, but this is because it is close enough to our galaxy to feel its attraction. The usual thing is for galaxies to move away from each other.

One of the first theories of the creation of the universe and its subsequent expansion was put forward in 1927 by a Belgian astronomer, the Abbé Georges Edouard Lemaître (1894–1966). Even before there was much evidence for an expanding universe, he deduced it from Einstein's theory of relativity. Later, after the idea had been strongly supported by the observations of Hubble and others, Lemaître developed his ideas, and suggested that the universe was originally small and highly compressed. Today this is the most popular theory of the origin of the universe, the big bang theory. This name was first given to it scornfully by the British astrophysicist Fred Hoyle, but it proved popular and has stuck. Other theories have been proposed, but all have been rejected by the great majority of astronomers.

There are excellent books about the big bang.[4] However, we can get the general idea of the theory and its validity even if we don't know much physics. Most of what we need was covered in chapter 3 of this book. We must remember that all of the chemical elements can be viewed as made up of three basic particles: electrons, protons, and neutrons. There are also four types of force that control the behavior of matter. Two of them we are already familiar with:

1. The *electrical* force, which, for example, keeps the electrons in their atomic orbits. It is a force of attraction when the charges are opposite (e.g., for atomic nuclei and electrons), and a force of repulsion when they are the same.
2. The force of *gravity*. This is by far the weakest of the four interactions. For two bodies, like a proton and an electron, that are oppositely charged, the electrical forces are about 10^{36} times greater than the gravitational attraction. However, gravity is of great importance when we consider interactions between large bodies like stars and planets.

The two other important forces in connection with atomic nuclei are:

3. The *strong nuclear* force, which binds protons and neutrons in the nuclei. This interaction has only a short range; a proton will only attract a neutron that is right beside it. In atomic nuclei, this force is roughly a hundred times stronger than the repulsions between neighboring protons; this explains how it is possible for the nucleus of the uranium atom (the largest reasonably stable nucleus) to contain as many as ninety-two protons, quite close to one another.
4. The *weak nuclear* force, concerned in subtle ways in radioactive decay. Its strength is about one hundred million millionth (10^{-14}) that of the strong force. It is, nevertheless, much stronger than gravity, by a factor of about 10^{25}. The weak interaction causes an isolated neutron to turn into a proton by emitting an electron.

In very approximate figures, the relative strengths of the four forces are as follows:

1. The force of *gravity*   1
2. The *weak nuclear* force 10^{25}
3. The *electromagnetic* force 10^{37}
4. The *strong nuclear* force 10^{39}

To understand the big bang theory we need to know a little physics relating to temperature effects. The big bang must have been accompanied by enormously high temperatures. Initial temperatures of over one billion billion kelvin have been estimated, but the exact value does not matter to the argument. Within a tiny fraction of a second from the big bang there was a great expansion, with a considerable temperature drop. Above the temperature of 3,000 K, electrons could not become associated with atomic nuclei, but freely pervaded the universe. The drop from the enormous initial temperature to 3,000 K occurred over a period of about 500,000 years, and during that time the photons (the particles of radiation) could not travel freely because of the electrons in their way; in other words, the universe was opaque to radiation. After the temperature fell below about 3,000 K, however, the situation greatly changed. The universe became transparent and light traveled freely. Electrons became associated with nuclei, and molecules could be formed by atoms uniting together in a variety of ways.

Overwhelmingly strong support for the big bang theory has been provided by two completely different lines of evidence. The first of these is *cosmic microwave background radiation*, which was discovered quite by chance. In the 1960s, American astrophysicists Arno Allan Penzias (b. 1933) and Robert Woodrow Wilson (b. 1936) of the Bell Laboratories in Holmdel, New Jersey, were exploring the Milky Way with a radio telescope, with the idea of improving communications with satellites. At a wavelength of about 7 cm, in the microwave region of the spectrum (see figure 16 on page 86), they found a greater effect than they could account for. The radiation was equally strong in all directions, and it even came from apparently empty sky. From the characteristics of the radiation, they were able to establish that it was emitted from a source that had an apparent temperature of about 3 K, that is, of about - 270 °C or – 454 °F.[5] They excluded the possibility that it was from some source on Earth. They also chased away two pigeons that had regularly deposited a white dielectric material on the antenna, but the pigeons came back and had to be removed more decisively.

None of their precautions removed the radiation, and it took some time to identify its source. It was finally concluded—and no better explanation has been suggested—that it is a kind of afterglow or "fossil" radiation resulting from the big bang. What presumably happened in the early stages is that as the universe cooled down, the strong forces first came into play, and these caused atomic nuclei to form. At first the universe was opaque to radiation, but after about half a million years it had cooled to about 3,000 K. The electrons then became associated with protons and neutrons to form atoms, and the universe became transparent to radiation. The temperature of 3 K for the background radiation tells us that, from the time it became transparent until now, the universe has expanded by a factor of about 1,000, because the temperature dropped by that factor, from 3,000 K to 3 K. Also, as a result of the stretching of space and in accordance with relativity theory, the wavelengths have increased by the same factor. Penzias and Wilson shared the Nobel Prize in physics for 1978 for this discovery of background radiation.

Spurred by their original discovery, many further investigations have been made of the background radiation, giving much more evidence for the big bang. Important data came from the Cosmic Background Explorer (COBE) satellite, launched into Earth's orbit in November 1989. A recorder measured the intensity of the microwave radiation over a range of wavelengths and in various directions. The results agreed exactly with the predictions of the big bang theory—so well that the average temperature of the radiation can now be given much more precisely as 2.736 K. There has even been an additional and particularly impressive bonus from these measurements. On the basis of the big bang theory, it was concluded that when the universe had cooled to 3,000 K, there would have been some density fluctuations; certain regions, where galaxies later formed, would have been of higher density than the rest. Because of these fluctuations, there would be regions of space where the temperatures would be slightly different, only one part in about 100,000. These tiny fluctuations over the sky were actually observed in the COBE experiments, exactly as predicted. It seems that the background radiation is indeed fossil radiation from the big bang.

The second line of evidence for the big bang theory is that it leads to a satisfactory theory of how the various atomic nuclei were formed. This formation of nuclei is known as *nucleosynthesis,* and if we understand in some detail how the nuclei are formed, we will have an explanation for the distribution of elements in the universe. We have seen that all atoms are composed according to a particular pattern. Each one has a tiny nucleus with electrons orbiting around it, the number of orbital electrons is equal to the positive charge on the nucleus, and this number defines the identity of the element. The nucleus is different for each chemical element, but can always be regarded as consisting of protons and neutrons. There is strong evidence that the atoms in a particular chemical element are, and always have been, the same throughout the universe. This evidence is provided by spectroscopic measurements made on many atoms, on Earth and in outer space. For example, the spectrum of a carbon atom in a galaxy that is ten billion light-years away, although as detailed as a fingerprint, is exactly the same as that of a carbon atom on Earth. That means that carbon atoms ten billion light-years away from us, which we see as they were ten billion years ago, were just the same as they now are on the Earth. The same is true of all of the chemical elements studied in this way.

About ninety reasonably stable chemical elements exist, and six of them play a particularly important role in the universe: hydrogen, helium, oxygen, nitrogen, carbon, and phosphorus. The first two have special status, since they are by far the most abundant elements. Of all the matter in the universe, about three-quarters (by weight) is hydrogen, and one quarter is helium. (A helium atom is four times as heavy as a hydrogen atom, so about 92 percent of all the atoms in the universe are hydrogen atoms and about 7.5 percent are helium atoms). All the rest of the elements together comprise only between 1 and 2 percent by weight. Although hydrogen is plentiful on Earth, it hardly exists at all in the uncombined state, as hydrogen gas, H_2. Because of its very low density, any free hydrogen gas would soon float to the top of the atmosphere and disappear into space. The same is true of helium gas, which exists only in the neighborhood of radioactive substances, which produce it as they disintegrate. There is an important difference between hydrogen and helium, however. Helium is highly inert, in the sense that it forms no compounds. Hydrogen, on the other hand, readily forms chemical compounds, many of which are abundant on Earth. Water, H_2O, of course, is very widespread on Earth, and many other hydrogen-containing compounds occur in, for example, fossil fuels and all animals and plants. Thus, in spite of not being able to remain on Earth in its uncombined state, hydrogen is abundant on Earth in its combined forms. Helium, which forms no compounds, is much less abundant, existing mainly around radioactive substances that produce it.

Helium thus plays a limited role on the earth, and indeed, was not discovered until the latter part of the nineteenth century. It is, however, used in many applications, such as in balloons and for investigative work. Hydrogen, oxygen, nitrogen, carbon, and phosphorus, on the other hand, are of great importance. Carbon has some particularly important and unique properties relative to life. Carbon atoms are able to link together in chains of indefinite length, forming molecules of great size and complexity, such as proteins and DNA. That is why the number of chemical compounds that contain carbon (called organic compounds) is enormously larger than the total number made up from the rest of the elements; we call these inorganic compounds.

One of the great triumphs of the big bang theory is that it gives a satisfactory explanation for the distribution of elements in the universe, for hydrogen and helium being the most abundant. The obvious explanation is that hydrogen was formed first, followed by helium, and the rest of the elements were formed later from these two elements. The first satisfactory explanation of how helium was formed was proposed by Russian-American physicist George Gamow (1904–1968) and his colleagues. Gamow was a large, enthusiastic man with an uncontrollable sense of humor, and in publishing his suggestion he perpetrated a joke that has become a scientific classic. The research was actually done with his student

Ralph Alpher (b. 1921), but Gamow considered that to publish with Alpher alone "seemed unfair to the Greek alphabet." He therefore added to it the name of Professor Hans Bethe (b. 1906), who had already made important contributions to the theory of nuclear reactions in the stars. The paper thus appeared in 1948 under the authorship of Alpher, Bethe, and Gamow. The fact that the date of the publication was the first of April was a particular delight to Gamow. I remember meeting Gamow a few times, and was always impressed by the fact that he spoke many languages with great fluency. The odd thing, however, was that each language as he spoke it sounded exactly the same to me.

Incidentally, one detail about the Alpher-Bethe-Gamow paper remains obscure to this day, since both Gamow and Bethe were reticent about it. One story is that before the paper was published, Gamow sent Bethe a draft of it asking if he would agree to his name being added, and that Bethe had no objection. The other is that Bethe did not know that his name would be added, and that at first he was not at all pleased, but later took the matter in good humor. Either way, it makes a good story.

The essence of the theory is as follows. First, protons became associated with neutrons, to form deuterium (see figure 8, on page 65), which they do readily at the enormous temperatures that existed in the early stages of the universe. Deuterium atoms, however, are uncommon—much less common than helium atoms—because the high temperatures of the early universe lead them to unite rapidly to form helium.[6] We have already discussed in chapter 3 the fusion reaction in which two deuterium nuclei form a helium nucleus (two protons and two neutrons):

$$\,^2_1H + \,^2_1H \rightarrow \,^4_2He + 23.7 \text{ MeV}$$

At the high temperatures that existed shortly after the big bang this process occurred rapidly, driven by the large amount of energy generated in the fusion process. Practically every deuterium nucleus that formed immediately after the big bang was converted into a helium nucleus. It can be deduced from the big bang theory and from our understanding of nuclear physics just why the amount of hydrogen in the universe is almost exactly three times the amount of helium, another very convincing argument in favor of the theory.

Hydrogen and helium are the only elements that could exist in the earliest stages of the formation of the universe, because the temperatures were so high that all other elements were unstable. The rest of the elements were made in the stars, where temperatures are lower, but still high enough for fusion reactions to occur. The formation of carbon presented a scientific problem that was finally resolved by the British cosmologist Sir Fred Hoyle (1915–2001). A carbon nucleus has six protons and six neutrons, and it could be formed from three helium nuclei, each of which has two protons and two neutrons:

$$\text{}^4_2\text{He} + \text{}^4_2\text{He} + \text{}^4_2\text{He} \rightarrow \text{}^{12}_6\text{C}$$

However, three helium nuclei are unlikely to collide, even in the dense atmosphere of a star. It is much more likely that the process occured in two stages, two helium nuclei first forming a nucleus of four protons and four neutrons, a nucleus of the element beryllium:

$$\text{}^4_2\text{He} + \text{}^4_2\text{He} \rightarrow \text{}^8_4\text{Be}$$

There seemed to be a serious snag, however. The $\text{}^8_4\text{Be}$ nucleus has never been detected, even in unstable form; the form $\text{}^7_4\text{Be}$, with only three neutrons, is the stable form of the element. Normally the light elements are most stable with an equal number of protons and neutrons. Beryllium is an exception to the rule. In fact, no isotope containing five or eight nuclear particles exists in any form, stable or unstable. This is rather curious; nuclei with every other number of particles up to over two hundred have been detected, although some of them are highly unstable. The absence of nuclei containing five or eight particles has to be taken into account when considering the synthesis of elements in the early universe.

Even if a $\text{}^8_4\text{Be}$ nucleus formed from two helium nuclei, there was little chance it would exist long enough to combine with another deuterium atom to form carbon by the process:

$$\text{}^8_4\text{Be} + \text{}^4_2\text{He} \rightarrow \text{}^{12}_6\text{C}$$

Hoyle suggested, however, that there was a way out of the difficulty if the $\text{}^8_4\text{Be}$ and $\text{}^4_2\text{He}$ nuclei came together in this reaction with an energy corresponding to an energy level in the carbon nucleus. Then their combination to form the carbon atom would be much more likely, and might take place before the unstable $\text{}^8_4\text{Be}$ atom had time to disintegrate. This speeding up of processes due to a coincidence of energies is called "resonance." When Hoyle made this suggestion, the energy levels of the carbon atom had not been measured, and so he arranged for the necessary experiments. He found that there really was a suitable energy level that allowed the process to happen efficiently. The formation of carbon in this way, in the stars, is therefore quite plausible. It is interesting that the proof of this possibility, which strongly favors the big bang theory, was obtained by Hoyle, who had always preferred a "steady-state" (no beginning or end) theory of the universe. According to this theory, as the universe expands, new matter is created to maintain the average density. New microwave evidence, however, and the interpretation of the distribution of the chemical elements, has left the steady-state theory with hardly any supporters.

Elements that are heavier than carbon were made in the stars in similar ways, by fusion processes. Those processes were suggested in a 1957 paper entitled "Synthesis of elements in stars," by Margaret Burbidge (b. 1922), her husband Geoffrey Burbidge (b. 1925), William Alfred Fowler (1911–1995), and Fred Hoyle. The paper is known to astronomers as B^{2}FH and it remains one of the classic papers of science, still fundamentally sound. Fowler, who for many years was a professor at the California Institute of Technology, made many important contributions to the understanding of stellar evolution and nucleosynthesis, and was awarded a Nobel Prize for physics in 1983. It may seem surprising that Fred Hoyle did not also receive a Nobel prize, since his contributions to the field were so considerable. It may have been due to Hoyle's support for the steady-state theory, which the latest work has made increasingly improbable.

There is one feature of nucleosynthesis that is of special interest and significance. One isotope of iron, iron-56 or $^{56}_{26}$Fe, is energetically more stable than many lighter nuclei and all of those that are heavier. If we construct an energy diagram of the nuclei of the elements, lined up according to their masses, iron is therefore at a minimum. For this reason, the nucleosynthetic processes that occur according to the B^{2}FH theory cause iron to be formed more easily, therefore in larger amounts, than many lighter and all heavier elements. The relatively higher abundance of iron in the universe therefore gives further strong support to the B^{2}FH theory of how the elements were formed, and to the big bang theory.

Nuclear reactions provide stars with their energy. Most of the elements produced in the stars remain there, but occasionally a star explodes at the end of its life cycle and becomes a *supernova*. As it does so, it expels some of the elements it contains. Aside from this, there is a steady flow of material issuing from stars; we know this from studies of the solar wind issuing from our Sun.

After the universe was first created by the big bang an enormous mass of gas expanded at high speed, accompanied, like all big explosions, by powerful shock waves. Little was present at first except hydrogen and helium gases, the latter formed by nuclear processes that were only present at the very beginning. From time to time, a massive cloud of gas would suddenly shrink as a result of gravitational attraction and form a dense region of spinning matter at a very high temperature. The oldest galaxies and stars, which contain the lowest proportions of the heavier elements, were almost certainly formed this way. Over the course of time, as matter was thrown out by supernovae, the gas gradually became more and more enriched with the heavier elements. Our galaxy was probably created about ten billion years ago. Initially it was swirling gas and dust, which became disk-shaped.

Stars, some with planets orbiting around them, were continually formed in it. About five billion years ago our Sun and its surrounding planetary system condensed from swirling gas and dust. The lighter material surrounding our Sun floated to the top, and eventually some of it condensed into the planet Jupiter, which is composed of gas at high pressure and probably has a solid core of rock and ice. The planets Neptune, Saturn, and Uranus were probably formed in a similar way. All of these planets have disk-shaped systems of rings; the nature of Saturn's rings, the most complete, was mentioned in chapter 3. These four planets are the outer planets of our solar system.

The heavier elements in the Sun became concentrated in the inner regions of the solar system and condensed into the inner planets, including the earth. Our planet, like the others, was thus created from the discarded material from stars that went through their life cycles between about five and ten billion years ago. Because of the way it was formed, Earth is significantly enriched in the heavier elements that are necessary for the formation of rocks. In particular, it was created with enough carbon, hydrogen in the form of water, and a few other elements, for life to become possible.

At first the earth was molten, and existed in that state for many millions of years before it solidified. Evidence from the rocks on the earth has led to conclusions about its age and suggests how it came to have its present form.

The age of Earth was a matter of considerable controversy for many years. We now know it to be a little more than 4.5 billion years, but until the end of the nineteenth century many investigators were arguing about much shorter ages. In fact, until about 1750, most people assumed that the Old Testament version was essentially correct. They thought the earth was a few thousand years old, and that all the sedimentary rock was deposited during Noah's great flood. Other surface features of the earth were assumed to have resulted from other catastrophes that occurred from time to time.

As geological observations were made, however, many investigators realized that such short periods were inconsistent with the evidence. They were forced to conclude that millions of years were necessary rather than a few thousand. One of the first proponents of the longer periods was Scottish geologist James Hutton, some of whose work was mentioned earlier. He became convinced that changes to the surface of the earth came about gradually over enormous periods of time. He realized that rivers play an important role in forming valleys and in other ways. They carry sediment into the sea where it accumulates in deposits, which might form new rocks by the action of heat from within the earth. He concluded that many millions of years, perhaps hundreds of millions, would be needed for mountains to form and erode.

Charles Darwin, who made numerous geological and biological observations during his famous voyage on HMS *Beagle* in the 1830s, also estimated that geological events on the earth occurred hundreds of millions of years ago. He was particularly interested in a wide valley known as the Weald of Kent in southeast England. Darwin concluded that this valley had arisen from the encroachment of the sea on the line of chalk cliffs on the south coast of Kent. He estimated that this process would occur at the rate of one inch per century, and from the configuration of the valley he concluded that it must have been formed 300 million years ago.[7] Other geologists, such as Sir Charles Lyell, also made estimates from the same kind of evidence, and from the speed with which sedimentary rocks might have been formed. His estimates, and those of many other contemporary geologists, agreed with Darwin's.

We now know that the age of the earth is at least ten times greater than these estimates, but geologists were strongly criticized by those who believed the estimates from the scriptures. There was also strong disagreement from an unexpected quarter, namely from the distinguished physicist William Thomson (later Lord Kelvin). He calculated the time it would take for the earth to cool to its present temperature from a molten state. He took into careful consideration the luminosity of the Sun, the expected cooling rate of the earth, the formation of the earth's solid crust, and even the effect of lunar tides on the rate of rotation of the earth. His initial conclusion, advanced in 1862, was that the earth could not have solidified more than 100 million years ago. In later publications and until about 1899 he proposed even shorter times, a few tens of millions of years.

Kelvin's estimate is still much more than suggested in religious writings, but less than most geologists had believed to be correct. Consideration was given to whether the geological events could have taken place much faster than previously thought possible. Many geologists felt sufficiently sure of their own conclusions that they ignored Kelvin's estimate, assuming that there was another source of energy in the earth that kept it warm for much longer periods. Kelvin himself had pointed out this possibility as early as 1862, referring to "sources of heat—now unknown to us . . . in the great storehouse of creation."

This additional source of heat turned out to be the solution to the difficulty. Soon after radioactivity was discovered in 1897, it became apparent that there was a vast store of radioactive substances in the earth. Radioactive disintegrations emit heat, which kept the earth warm for a longer period. Kelvin's upper limit of 100 million years was thus far too low. Estimates were made factoring in the heating effect of radioactive material discovered in the earth, and these were consistent with the longer periods that the geological evidence required for the age of the earth.

The study of radioactive substances also contributed more directly to reliable estimates of the age of the earth. Earlier we saw several examples of radioactive processes, and some of these can be used to arrive at reliable estimates of the ages of rocks and other materials. One process is the disintegration of uranium-238, the isotope present in the largest proportion (about 99.3 percent) in natural uranium:

$$^{238}_{92}U \rightarrow {}^{4}_{2}He \text{ (an } \alpha \text{ particle) } + {}^{234}_{90}Th$$

The half life of the process is about 4.5 billion years, which makes it particularly convenient for measuring the ages of rocks.

One way of estimating the age of a rock is based on the fact that, as uranium disintegrates, it produces helium gas, which is chemically inert and stable. Helium easily leaks out of molten rock, but after the rock cools and solidifies it remains trapped. Analyzing a sample of rock for the amounts of uranium and trapped helium will show how much time has elapsed since the rock solidified. Experiments of this kind have been done many times, and show that the earth has existed in its present solid state for about 4.5 billion years. Some rocks are indeed formed at later times, but no rocks have been less than several hundred million years. Therefore, entirely independent measurements showed that the estimates of the geologists were by no means too high. The values deduced from religious writings are completely wrong.

Experiments determining the amounts of helium present in uranium minerals were carried out by Ernest Rutherford, who also did pioneering work on radioactivity and atomic nuclei. He obtained ages of about 500 million years for the particular rock samples he studied, and thus proved conclusively that Lord Kelvin was wrong. In 1904 Rutherford presented an important lecture at the Royal Institution in London on the age of the earth and the production of heat by radioactive substances.[8] Entering the lecture room, he was disconcerted to see that Kelvin, then age eighty, was in the audience. However, Kelvin drifted off to sleep, but Rutherford reported later that when he began to speak about the age of the earth he saw "the old bird sit up, open an eye and cock a baleful glance" at him. With great presence of mind, Rutherford modified his lecture somewhat, reminding his audience that in 1862 Kelvin had referred to "sources of heat now unknown to us," and that he had qualified his estimate of the age of the earth by writing "provided no new source of heat is discovered." Thus, said Rutherford with admirable tact, "the audience must admire the foresight, almost amounting to prophesy, which had made Lord Kelvin so qualify his calculations." Rutherford was then relieved to see that "the old boy beamed" at him.

In the last chapter we briefly met British geologist Arthur Holmes in connection with early work on continental drift. Holmes was also a great pioneer in determining geological age by the radioactivity methods. In

1913, when he was only twenty-three years old, he published a book, *The Age of the Earth*, which was soon regarded as a scientific classic. It contained the passage, "It is perhaps a little indelicate to ask our Mother Earth her age, but Science acknowledges no shame and from time to time has boldly attempted to wrest from her a secret which is proverbially well guarded." The book gave an excellent account of the results obtained by radioactivity techniques and of how they relate to more conventional geological studies on rates of erosion and sedimentation. Over the years the radioactive work has been greatly extended, but without requiring any important changes to Holmes's early conclusions.

Uranium disintegration can be used to determine the ages of rocks in another way. The $^{234}_{90}$Th nucleus that is a product of the disintegration is itself radioactive, and the reaction on page 148 is only the first step in a chain of radioactive processes, all of which occur more rapidly than the original step. The final product is lead-206, and we can write the whole chain of processes as:

$$^{238}_{92}U \rightarrow {}^{206}_{82}Pb + \text{several particles}$$

Since the first step, with a half-life of 4.5 billion years, is the slowest in the chain, it controls the overall rate, so we can regard the half-life of the whole chain to be 4.5 billion years.

Suppose that while a planet or a meteor thrown out from the Sun was still molten, some uranium was trapped in it. At first the uranium could move fairly freely in the liquid rock, but after the rock solidified it would be trapped, and the lead formed would then be forced to remain near to the uranium. Then, from the ratio of the amounts of uranium-238 and lead-206 found in any part of the rock, we can calculate the time that has elapsed since the rock solidified. This technique has been applied to a number of meteorites found on the earth. The meteorites classified as "stony" are the oldest, having an age of about 4.6 billion years. They are the oldest known objects in the solar system, so that at least some parts of the solar system must have solidified 4.6 billion years ago. Rocks that were brought back to Earth from the Moon by the Apollo astronauts are a similar age.

Several other radioactive substances have been used to determine the ages of rocks, and one example is the rubidium-strontium method. Rubidium, Rb, is a fairly uncommon metal, similar in its properties to the more common sodium and potassium, both of which are widely distributed in minerals. Rubidium forms no minerals of its own, but is so like potassium that it can substitute for potassium in all potassium-containing minerals. It occurs in small but easily detectable amounts in several common minerals, such as the micas and certain clay minerals.

Rubidium has two naturally occurring isotopes, $^{85}_{37}$Rb and $^{87}_{37}$Rb, whose relative abundances are 72.2 percent and 27.8 percent, respectively. The latter is radioactive, emitting a beta particle to give an isotope of strontium, $^{87}_{38}$Sr:

$$^{87}_{37}\text{Rb} \rightarrow {}^{87}_{38}\text{Sr} + {}^{0}_{-1}\beta \text{ (a beta particle)}$$

Since strontium has properties that are similar to calcium, it can replace it in many minerals. Strontium has four naturally occurring isotopes, the particular isotope $^{87}_{38}$Sr normally occurring 7.04 percent of the time. However, when strontium occurs in the neighborhood of rubidium-containing rocks, its percentage is higher. Measurements of the relative amounts of $^{87}_{35}$Rb and $^{87}_{37}$Sr in such circumstances lead to estimates of the age at which the rocks solidified. The results are always consistent with those obtained in other ways, giving ages of several billion years for the oldest rocks, with shorter ages for some metamorphic rocks that later underwent a fiery transformation.

Today the science of isotope geology is highly developed, leading to a coherent picture of the ages of various geological events. The circumstantial evidence from isotope geology that the earth is at least 4.5 billion years old is overwhelming. Added to all the other evidence for this number, it is hard to see how anyone could dispute it.

*   *   *

Life appeared at least 3.5 billion years ago, about a billion years after the earth was created. The evidence for this age is the ancient remains of living organisms, whose age can be obtained by radioactive dating. Various methods are used, all leading to the same answers.

One method is to use potassium-40, which decays into argon and has a half-life of 1.3 billion years. The potassium-argon ratio method is often used to date early fossil remains. Another method, using radioactive carbon, is particularly useful for estimating the age of more recent material such as wood and bones; this technique is known as *carbon dating*. Carbon-14 has a half-life of 5,730 years and decays to nitrogen-14 and a beta particle:

$$^{14}_{6}\text{C} \rightarrow {}^{14}_{7}\text{N} + {}^{0}_{-1}\beta$$

This carbon-14 isotope hardly exists at all on the earth's crust, but it is constantly produced in tiny amounts in the atmosphere by the action of cosmic radiation on nitrogen molecules. The neutrons in the cosmic rays convert nitrogen into carbon:

$$^{1}_{0}\text{n} + {}^{14}_{7}\text{N} \rightarrow {}^{14}_{6}\text{C} + {}^{1}_{1}\text{H}$$

Sunlight causes some carbon-14 to combine with oxygen in the air to form carbon dioxide (CO_2), which enters into all living organisms. As long as the organism is alive, the amount of carbon-14 it contains is maintained by atmospheric carbon-14, which can enter the system freely. When the organism dies, carbon-14 ceases to enter the organism, and the carbon-14 already present begins to decay. The amount of carbon-14 that remains in a fossil enables scientists to determine when the organism died.

The oldest fossilized organisms discovered so far are in rock formations in Western Australia. In the Pilbara area, about forty kilometers west of the small town of Marble Bar, John Dunlop, a geology student, discovered in 1980 some stromatolites. These are structures that are deposited, layer by layer, by cyanobacteria, a form of bacteria that, like green plants, use photosynthesis—the process by which water and carbon dioxide are converted into organic compounds. Radioactive dating gave the age of these stromatolites as 3.5 billion years. If photosynthesis, a somewhat complex process, had already evolved, then living systems must have originated a good deal earlier than 3.5 billion years ago.

Since Dunlop's original discovery, many other signs of very early life have been detected. Some of these were found near Darwin, in Western Australia, named after Charles Darwin, who worked in that region during his trip on the *Beagle*. He was surprised to find no fossils, but they are there, just too small to be detected without modern techniques.

In the 1990s, evidence for early life was also found in a remote mountainous region at the edge of the massive Greenland ice sheet. An expedition led by Gustav Arrhenius of the Scripps Institute of Oceanography in California discovered traces of carbon-containing deposits in the form of tiny grains, laid down by primitive living organisms. The evidence that they were once alive comes from a particularly sensitive radioactivity technique developed recently. Carbon-12 is the most common isotope of carbon, the nucleus of its atom containing six protons and six neutrons. A small proportion (about 1.10 percent) of the carbon in nature occurs as carbon-13, which has an extra neutron in the nucleus. Because of its smaller mass, carbon-12 chemically reacts slightly faster than carbon-13, but otherwise the chemical properties of the two forms are almost identical. With this slightly greater rate of reaction, more carbon-12 accumulates in living organisms than carbon-13, and the excess gives an indication of age. In the Greenland rocks, the carbon-12 was about 1 percent higher than normal, which is confirmation that the material was really of living origin. The age was about 3.85 billion years.

It is impossible to be certain how the earliest living organisms came into being. For some time it was thought that protein molecules played the essential role in living systems, and various attempts were made to see whether, in the kind of atmosphere believed to exist on Earth in those early times, they might have formed naturally from available substances. In 1953 experiments were carried out by Stanley Miller (b. 1930), a graduate student in the laboratories of Harold Clayton Urey (1893–1981), who had won the 1934 Nobel prize for chemistry and was distinguished for his work on isotopes and the distribution of chemical elements. Miller, with Urey's rather grudging approval at first, made mixtures of simple chemicals thought to be present on Earth 3.5 billion years ago, and subjected them to electric discharges that simulated lightning. He found, for one, that if water vapor, hydrogen, methane, and ammonia were held at 100 °C (212 °F), and subjected to electric discharges for a week, some amino acids and more complex substances were formed. Since protein molecules are produced by the combination of amino acids, these experiments seemed to allow the possibility that the first life forms had arisen in this way.

However, as a result of Crick and Watson's determination of the structure of DNA and later studies (see chapter 4), we know that the key substances in living systems are not the proteins but the DNA molecules, which are directly concerned in replication. DNA is much less likely to form by chance than protein, since DNA is a much more complicated molecule than a protein. Of course, the production of a molecule of DNA, which must have happened at some stage, may not have been the first formation of a self-replicating substance. Graham Cairns-Smith of the University of Glasgow has suggested that inorganic clay crystals may have been the first replicators, and DNA came later.

It seems impossible to learn how life originated on Earth, but it is interesting to speculate on whether there is a reasonable probability of its happening by chance. Several facts have to be taken into account. One is that life apparently began at least 3.5 billion years ago, which is about a billion years after our Earth had cooled enough to make life possible. Also, astronomers have estimated that there are about 100 billion billion (10^{20}) planets, and it is possible that some of these could sustain life. If life had not begun on this planet it might well have begun on another. However, most scientists consider it unlikely that molecules as complex as DNA formed by accident, though they could have been produced from simpler molecules, such as those considered by Cairns-Smith. It is difficult, however, to quarrel with the view expressed by Sir James Jeans at the beginning of this chapter. Most of the universe is highly inhospitable to life.

It is usually assumed that our ultimate ancestor originated on Earth, but primitive life might have begun elsewhere and been brought to Earth, perhaps by a meteorite. The possibility that it originated on Mars is explored in an interesting way by Paul Davies in his book, *The Fifth Miracle: The Search for the Origin and Meaning of Life*, published in 1999.[9] Mars is now inhospitable to any form of life, but there is evidence that it once had lakes and rivers, and an atmosphere more like our own. It is therefore possible that life in the universe began on Mars and was conveyed to Earth by a meteorite. Small samples brought back from Mars in 1996 were said by some to indicate the possibility of life but, as Davies makes clear, the evidence is far from conclusive.

In whatever way the initial replicating molecule was formed, we can hardly doubt, from the evidence, that all living things now on this Earth are descendants of a microorganism formed about 4 billion years ago. One reason for believing this relates to the genetic code and the laws of probability. We have seen in chapter 4 that we can think of the genetic code as a tiny dictionary with sixty-four codons; this is the number of possible ways the four bases T, C, A, and G can be arranged in triplets—such as CGC, the code word for the amino acid arginine. These sixty-four codons are the codes for the twenty amino acids plus a punctuation mark; it is as if the sixty-four words of one very primitive language were translated into an even more primitive one in which there are only twenty-one words (twenty acids plus a punctuation mark). The odds of getting the same 64:21 coding twice by chance can be calculated to be less than one in 10^{30}. It is thus practically impossible that the different species of bacteria, plants, and animals were created separately.

Another good reason for thinking that all life had a common ancestor is that all of the molecules that play essential roles in living systems have the same *chirality* or "handedness." If you look at your hands, you see that they are different, but are similar in the sense that one hand is more or less the mirror image of the other. The same is true of the amino acids. A given amino acid can exist in either of two forms, the D form and the L form; each is the mirror image of the other, but because of a particular feature in their structures, one of them cannot be superimposed on the other. It turns out that in every living system that has ever been studied, all the amino acids are in the L form. Moreover, the D forms cannot be accommodated into living systems. If, for example, someone gave us a piece of meat that had been synthesized, but had every amino acid in its protein molecules in their D forms, it might look exactly the same as meat. If we tried to eat it, however, the enzymes in our stomach and intestines would be quite unable to digest it.

All organic molecules in living systems, including the proteins, the DNA, and the RNA, contain only L molecules. A form of life using all D molecules, or containing various mixtures of L and D forms, would have been just as possible, but has never been observed. Perhaps life did originate several times and in different ways, but if so, only one arrangement, all L, has survived. If living things came from several origins, surely some of them would contain D rather than L molecules.

Over the ages, living organisms have adapted to increase their potential for survival and reproduction. As certain characteristics are transmitted from parent to offspring, and as mutations occur from time to time, some organisms prove to be poorly adapted and die out, while others increase in number. In accordance with Darwin's theory, there is survival of the fittest: the weakest organisms (they don't live long enough to reproduce) fail to survive, while the stronger ones continue the line.

Different species can be formed from mutations. They can also be formed when organisms of the same species become separated geographically. The separated species then develop in different ways, becoming different from each other after many generations as a result of natural selection, until finally they are so different that they can't interbreed. By definition, they are then separate species. For example, why did red squirrels and gray squirrels, although resembling one another in so many ways, become different species? Red squirrels are indigenous to Europe—the few in North America were transported there. Gray squirrels, sometime called Canadian squirrels, are indigenous to North America, although some were introduced into Europe (being squirrels, they propagated rapidly and are now numerous). From their great similarity we can conclude that long ago their ancestors belonged to one and the same species. It seems likely that many generations ago a group of squirrels became separated into two populations, perhaps by a mountain range or by water. Later, as the earth's configuration changed, they became separated by the Atlantic Ocean. They continued to breed, and over the generations, as a result of mutations and natural selection, the members of the two populations became sufficiently different from one another that they could no longer interbreed.

There are three main classes of living beings: bacteria, plants, and animals. In the early days of the theory of evolution, before much had been learned about genetic mechanisms, it was thought that evolution occurred by simple pathways, something like the following:

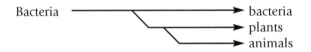

Once the first animals were formed, it was thought that there was a progression through monkeys, chimpanzees, Neanderthal man, and finally our species, *Homo sapiens*.[10] We now know that this is wrong. We did not descend from monkeys or chimpanzees, and not even from Neanderthal man. They belong to different branches of the tree of life, shown purely schematically in figure 23.

As Richard Dawkins has explained clearly in his fascinating book, *River out of Eden*, a good analogy for the evolutionary process is a river.[11] Unlike any real river, it is incredibly complicated; its source corresponds to the microorganism from which all life originated, and it forms a vast number

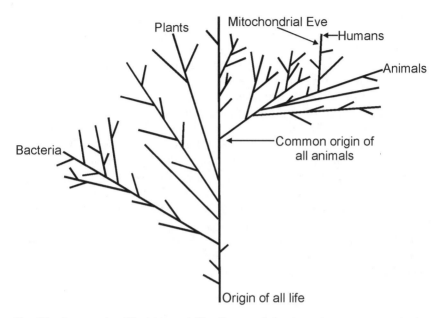

Fig. 23. A very simplified tree of life. Some of the branches never reach the tree's outer extremities and represent extinct species, such as dinosaurs and Neanderthal man. A more detailed (and enormously large) diagram would show thirty million branches corresponding to still-living species; ten times as many have not survived.

of tributaries, finally emerging at the sea (the present time) at 30 million places, since that is the number of species existing now. More often than not, tributaries dried up and never reached the sea, representing the many species, like dinosaurs, that became extinct.

The details of this "river out of Eden" are still being studied; after all, there are thirty million species to deal with. Modern molecular genetics has contributed greatly. For example, there is much evidence from conventional biology that apes (including gorillas and chimpanzees) are closely related to each other and to humans; the DNA differences between all of them only amount to 1 or 2 percent. It is inferred from this and other evidence that the common ancestor that we share with apes, gorillas, and chimpanzees existed four or five million years ago.

Interesting conclusions emerge when we go into further details of the genetic structures. An important protein called "cytochrome c" is made up of about a hundred amino acids, its formation in humans being determined by a string of 339 codons. A monkey, another close relative of ours, also has a cytochrome c, which differs from ours by only one amino acid out of the hundred or so. In the monkey, it has almost exactly the same properties and performs the same functions; the section in the monkey's genes that produces it differs from ours in only one codon out of the 339. This obviously gives overwhelming support to the idea that monkeys and humans came from a common ancestor in the relatively recent past, probably about four or five million years ago.

On the other hand, when we compare the cytochrome c from yeast with that from humans, 45 out of the 339 codons are different. Again, this makes sense because humans are very different from yeast, and the river leading to humans and yeast obviously split in a much more distant past—much more than four or five million years ago. The difference between the numbers for pigs and yeast is also forty-five. The river leading to pigs split from the one leading to us much more recently than their common river separated from the one leading to yeast. There are many relationships of this kind, all giving strong support to the theory of evolution.

With some of the finer details there has been some disagreement between experts, but in view of the complexity of the situation, and the vast number of genes involved, this is inevitable.

Chapter 6
Science and
Culture

Culture is roughly anything we do and the monkeys don't.

—Attributed to Lord Raglan (1788–1855)

O f all the branches of knowledge, science has changed the most over
the last few centuries. If Isaac Newton were to come back to earth, he
would need a long period of study to appreciate the present state of scientific knowledge. As to modern technology, Newton would be amazed and
probably at first intimidated by fast cars, television, computers, jet-pro-
pelled planes, and all the other devices with which we are familiar today.

By its very nature, science is subject to change and advancement. Nevertheless, the methods now used in all branches of knowledge are much
the same; there is no special scientific method, but a judicial or academic
method, characterized by the unbiased and critical processing of information. These are the methods that should be used in making all our decisions. Some of these relate to broad principles, or matters of public
policy—the kinds of things that are dealt with in sermons, newspaper edi-
torials, and letters to newspaper editors. As far as possible, we should make
sure we use rational arguments to form opinions, and not just prejudices
that we have mindlessly picked up from others. That is not to say that
absolutely everything we do should be determined completely by reason.
One feature that makes human beings different from any other animal
species is that our emotions and culture play a great part in our lives. By
culture, I do not mean anything that might in some quarters be regarded as

"elitist," but rather the kind of thing that Lord Raglan had in mind when he made the remark that appears at the beginning of this chapter. Culture comprises language, folk-music, myths, fashions in clothes, architecture, sports, philosophical beliefs, etc.

This chapter will discuss some of the more general matters on which we form opinions. I begin with scientific problems, because technological advances have such an enormous impact on our lives and culture. Science presents us with a special challenge, because it is difficult, even for a scientist, to keep sufficiently well informed about much of it. It is much more difficult for nonscientists. Decisions often have to be made by politicians and others who do not understand the science involved. Scientists are becoming increasingly concerned that the wrong decisions are too often made on matters of public policy, and this will cause great difficulties in the future. As Mark Twain once said, "It is difficult to make predictions, especially about the future." It is impossible to make reliable predictions of the technological events and needs of the future and, as shown in table 3, even distinguished scientists and technologists have often gone badly wrong.

Table 3: Some Bad Predictions

STATEMENT	SOURCE	YEAR
Invention		
Everything that can be invented has been invented.	Charles H. Duell, Commissioner, U.S. Patents	1899
Telephone		
This telephone . . . is inherently of no value to us.	Western Union internal memorandum	1876
It is certainly a wonderful instrument, although I suppose not likely to come to any practical use.	Lord Rayleigh	1876
Americans have need of this invention, but we have not. We have plenty of messenger boys.	William Preece, chief engineer of the British Post Office	1877

Radio and Television

Speech over the radio is as likely as a man jumping over the moon.	Thomas Edison, said to R. A. Fessenden, the world's first disk jockey	1899
I don't think this business of television is likely to come to much.	J. J. Thomson	1930

Aircraft

Heavier-than-air flying machines are impossible.	Lord Kelvin	1885
I have not the smallest amount of faith in aerial navigation other than ballooning.	Lord Rayleigh	1889
The gas turbine could hardly be considered a feasible application to airplanes.	U.S. National Academy of Sciences, Committee on Gas Turbines	1940

Nuclear Power

Anyone who expects a source of power from the transformation of [the nuclei of] atoms is talking moonshine.	Lord Rutherford (his opinion had changed a year or two later)	1933

Computers

I think that there is a world market for maybe five computers.	Thomas Watson	1943
There is no reason anyone would want a computer in their home.	Ken Olson, founder of Digital Equipment Corporation; chairman of IBM	1977

640 K ought to be enough for anybody.	Bill Gates	1981

Space Travel

The possibility of travel in space seems at present to appeal to schoolboys rather than scientists.	Sir George P. Thomson	1956
Space travel is utter bilge.	Sir Richard Woolley, Astronomer Royal	1956

Antibiotics

It is time to close the book on infectious disease.	U.S. Surgeon General	1969

Since the beginning of the nineteenth century in particular, science has had an increasingly important effect on the way we live. Life at the beginning of the twenty-first century is vastly different from what it was at the beginning of the nineteenth century. Travel was then by horse and carriage; now it is by car, train, and plane. Lighting was by oil lamps and candles, now it is by electricity; heating was by coal, now mainly by gas, oil, and electricity. Radio, television, and computers were quite unknown until the twentieth century. Medicine has been transformed from largely empirical procedures into a sophisticated practice that is heavily dependent on the basic concepts of molecular biology, physics, and chemistry.

Today most of us take these changes for granted. In their debates, politicians rarely take any explicit account of technological change, except to complain that it costs money. Governments usually fail to recognize that all modern technology is highly dependent on pure science, which is concerned with basic principles. As a result, governments usually do not give adequate financial support to pure science. In the fortunate regions of the world, people make full use of modern conveniences like an adequate supply of electric power, plentiful supplies of food and water, anesthetics, and pharmaceutical drugs. At the same time, paradoxically, some of them—and particularly the more vociferous—are apt to attack some of the scientific tools and technologies, such as nuclear power and genetic engineering, which have proved so useful in providing these advantages. Most scientists think that these techniques must continue to be used if facilities are to be in adequate supply over the years to come.

In view of these inconsistencies, it seems worthwhile to discuss the progress of science as it affects our lives, and the public's attitudes to it. Great changes were brought about from the world war of 1939 to 1945. The scientific successes of the war, such as the harnessing of nuclear energy, the application of radar, and the development of computer technology, produced at first a general public euphoria. Many thought that nuclear power would offer a complete solution to all problems of energy and food supply, and that communications techniques would greatly enhance dissemination of information. Science seemed to offer a general panacea, and there was a dramatic increase in the number of students wanting to enter scientific fields. One result was the founding of many new universities; another was an increase in government support of science.

From a purely technical point of view, these hopes have been realized. Nuclear power plants have been installed in a number of countries, and in some of them, such as France and Belgium, provide a substantial portion of energy. Improvements in communications technology have been spectacular, resulting mainly from the introduction of the transistor in 1947 and the microprocessor in 1971. These two technical innovations, by allowing extensive miniaturization, have completely transformed the way business is transacted and the everyday lives of hundreds of millions of people. They exert a great influence in areas such as the automobile industry, the manufacture of household devices, and the practice of medicine.

Chemists, biochemists, and medical researchers have also made important advances. New plastics, some with quite remarkable properties, have become available. Many medical needs have been satisfied by a variety of new pharmaceutical products, so that some diseases have been almost completely eradicated. Treatments for many other conditions have been discovered, much progress has been made in the understanding of genetics, and there is great promise for the future.

In spite of these technical successes, the public euphoria about science and its applications went into a deep recession in the early 1960s, mainly because some of the ill effects of science had become apparent. Technologies, if not properly controlled, can seriously harm our environment. Improvements in nuclear technology led to a proliferation of nuclear weapons, and to real and potential pollution. Improvements in communication made it easy for people to disseminate hateful material.

We must recognize two factors of fundamental importance. One is that, once scientific knowledge has been gained, there is no way it can be hidden or suppressed. The second is that science cannot advance by making only socially desirable innovations. Society at large must make responsible and informed decisions about the consequences and use of scientific and technological advances.

Public campaigns sometimes exert a strong influence on public policies. Sometimes they do some good, but too often they are based more on emotion than on accurate scientific information and sober decision-making. It is undeniable that there must be standards and safeguards to protect the health and welfare of the earth's inhabitants. They should, however, be imposed with all factors taken into account, not just some undesirable or risky consequences. Instead, these must be judiciously balanced against the advantages. Environmental spokesmen are most effective when they get their facts straight and base their conclusions on all of the evidence, not just the evidence that supports a particular case.

Physics has come to have a bad name in some people's minds, largely because it has given us the atomic bomb. Nuclear energy plants, using a related technology, have some disadvantages, and in some places there has certainly been lack of adequate control. However, it is wrong to condemn nuclear power plants on the sole basis of problems that have arisen with a few of them. The French in particular have been remarkably successful in using nuclear power with no serious problems. This is one field in which a great deal of nonsense is uttered by propagandists when they select the particular information that suits their cause. Chapter 1 contains a statement (see page 22) that the amount of energy required to mine and purify uranium for use in a nuclear plant is very great, with the implication that nuclear energy offers no advantages over conventional fuels like coal. It is true that it takes about a hundred times more energy to produce a ton of suitable uranium than a ton of coal, but a ton of uranium in a nuclear reactor produces 20,000 times more energy than a ton of coal in a furnace. Aside from the energy advantage, the production of uranium is much safer; compare the number of uranium miners killed with the number of coal miners killed. Also, the nuclear industry produces much less pollution than energy technology based on fossil fuels. The balance of this evidence is thus squarely in favor of nuclear energy.

Environmentalists sometimes tell us that our energy needs can be satisfied without using fossil fuels or nuclear power plants. Wind and solar power, and geothermal energy are certainly less polluting, and are valuable as supplementary sources of energy. Simple estimates show, however, that it is impractical for these sources to produce the large amounts of energy required by modern industry, energy that we expect to be available at all times, independent of sunshine and wind. Solar power may be adequate to keep a house warm in a temperate climate, but we need much more energy. To satisfy the energy needs of ten million people in a developed country would require solar reflectors totaling an area of several hundred square miles. Aside from the fact that the construction of such a plant would create massive atmospheric pollution, its presence would have devastating effects on the wildlife and on the weather. Since we know how much energy is emitted by the Sun, and have a detailed understanding of the thermodynamic and other efficiencies of collecting solar energy, we can be sure that technical improvements can never allow solar energy to be a practical solution to the energy problem. Wind energy, also derived from the Sun's energy, is also insufficient. The supply of oil and other fossil fuels is limited, and some experts predict that it may be depleted within a very few decades.[1] Whether we like it or not, it does seem that nuclear power will turn out to be the only way for the future.

Aside from the great importance of nuclear science in the production of energy, nuclear technology plays a large part in our daily lives in other ways. It sterilizes food and protects the quality of water. Many lives have been saved by its use in smoke detectors. Its applications in medicine are numerous and widespread. Nuclear science plays an important role in medical diagnosis and in therapeutic procedures for the treatment of health problems like infections, cancer, and heart disease. Nuclear medicine is less invasive than surgery, and, under some circumstances, the doses are too small to produce ill effects. In spite of all these valuable applications, there are some who advocate the complete abandonment of nuclear technology. Obviously they have not made a careful appraisal of the situation, balancing the advantages against the drawbacks.

Chemistry has also become an ugly word to some, associated with the pollution of the atmosphere, the drug culture, and the proliferation of weapons. Foolish decisions have sometimes been made—for example, to ban the use of chemicals such as chlorine—but these decisions create more problems than they solve. Some people talk as if they would like to suppress chemistry altogether. Such people would be indignant if they found themselves denied the use of gasoline, anesthetics, and pharmaceutical drugs. They forget that the work of chemists is essential to the production of these materials. There is no possible way chemistry can advance and make only socially desirable materials. It is for society to determine what use is made of chemicals.

A widespread misunderstanding about chemistry relates to whether substances have been extracted from materials found in nature or have been synthesized by chemists in the laboratory. People who hold extreme views on such matters have even invented a confusing and misleading terminology. They apply the words "natural" or "organic" when direct use is made of substances present in nature, while materials made in the laboratory are called "artificial." In reality, the situation is very different. Consider first what chemists call the chemical *elements*, of which carbon, iron, and nickel are examples. There are two ways we can obtain these elements. The ordinary and most common method is to get them from the earth or the earth's atmosphere. Iron and nickel, for example, are found in ores, which are treated by chemical methods to obtain the pure elements. In a few special cases we obtain elements by means of nuclear devices, which produce them from other elements, usually by bombarding them with neutrons. The radioactive isotopes that are extensively used in therapy are produced this way.

An element is an element no matter how we obtain it, and it makes no scientific sense to suggest that an element obtained in one way is superior to the same element obtained in another. Advertisers have sometimes tried to make the public believe that calcium produced from seashells is better than that produced from limestone. This is scientific nonsense. Of course, impurities left behind from an extraction process may have an effect, but not necessarily a beneficial one.

It is equally true that a chemical compound is the same wherever it came from. It makes no difference whether we have obtained it directly from a natural source (from a plant, for example) or have made it in a laboratory or commercial plant. There is one important difference between the two situations, however. Of all the millions of compounds that chemists have identified, only a tiny proportion exist in nature. If we insist on "natural" compounds, we unnecessarily deprive ourselves of an enormous number of compounds. The vast majority of anesthetics and pharmaceutical drugs cannot be extracted from materials in our environment. Many of them could not possibly be produced in adequate amounts except in a laboratory or plant. They are none the worse for being "artificial," and one does oneself no favor by refusing to use them. It may well be that in some cases the production of "artificial" substances has undesirable consequences, but these must be balanced against their advantages.

It thus makes no sense to argue that "natural" or "organic" methods are necessarily more desirable than "artificial" ones. In fact, materials obtained from "natural" products, as well as "artificial" substances, may have some undesirable impurities. The important point is that if synthetic methods of producing many materials were abandoned, their supply would be quite inadequate—if it were possible to find them at all.

Campaigns are often directed against the use of fertilizers made in the laboratory. However, if these methods were suppressed, a vast number of people would starve. Some chemical fertilizers may certainly have undesirable effects, but there are many that can be used satisfactorily. They have been used successfully for decades in many countries, including the United States and Britain, where health and longevity standards have improved. Those who oppose the use of chemicals suggest that the solution to the world's food shortages lies in "organic farming," which means farming with chemicals found in nature rather than man-made. Aside from the absurdity involved in the expression "organic farming," it has been conclusively established that the yields from "organic farming" are significantly lower than those obtained by the use of synthetic fertilizers. People should not argue against such practices unless they can suggest a more desirable way of avoiding extensive food shortages. If they cannot, there is surely a moral obligation to use synthetic fertilizers in order to feed its inhabitants.

Similar objections have been raised to the application of genetic engineering in farms, and there has been a great furor about the technique; there are even objections to *experimentation* on genetic manipulation. But genetic manipulation has been practiced by hit-and-miss methods for thousands of years. Experimenting with seeds, fruit-and-vegetable farmers have greatly improved the quality of many of their products. Tomato plants originally bearing pea-sized fruit have, by suitable manipulation, given rise to tasty fruit. Early varieties of corn producing tiny ears were transformed into varieties bearing large cobs.

Over the centuries the public have welcomed these genetic alterations because they have greatly enhanced the quality of food. Genetic advances have staved off hunger and starvation for millions of people. It is therefore paradoxical that, now that scientists have much better control over genetic manipulation because they know much more about the genes, some people become incensed and want a ban on the procedures. After all, over thousands of years the same plant has undergone extensive cultivation, with numerous, little-understood genetic alterations. This has happened without the control and testing required today for genetic engineering. Surely modern methods, with all their precautions, are better than the hit-and-miss procedures of the past.

Aside from the question of food, genetic engineering has shown enormous advantages in a variety of other ways. Some of these were mentioned at the end of chapter 4, and here is another example. Two hundred million people in the underdeveloped world suffer from vitamin A deficiency, causing seriously impaired sight and sometimes blindness. One way their problems can be alleviated is with genetically engineered rice with enhanced vitamin A. There has been opposition to such techniques, but there is no convincing evidence that they represent significant dangers; after all, controlled genetic changes are likely to be better than those that occur by chance as a result of mutations. Until better suggestions are made, it would be unreasonable to ban the use of genetic engineering when, as is often the case, the advantages are so significant and widespread.

Sometimes ethical objections are raised to genetic technology. Again, we must balance the advantages against the disadvantages. In some cases it seems to me that the advantages are so overwhelming that they override any possible ethical objections. Today many pharmacological substances, produced from genetically engineered bacteria and yeasts, have relieved suffering and saved lives. Even when human genes are used, it is difficult to see that such procedures are necessarily unethical.

Cloning is a type of genetic engineering that is widely misunderstood by the public. One cause for confusion is that cloning can mean several different things. It can mean taking the genome of an individual, even a human, and making an exact copy. Most people agree that such a procedure might be acceptable with species other than humans. Cloning can also mean cloning a particular organ, such as a heart or a liver. In the future it may be possible to clone a new heart from a patient's own genome and replace a defective one. Techniques of this kind may be commonplace in the future, and it is difficult to see that there could be any ethical objections in the case of a heart or a liver. On the other hand, if the cloning of a person's brain were ever to become feasible, would that be ethical? It would certainly make most of us uneasy, and we obviously need laws to control such procedures. But it is surely ignorant and foolish to condemn cloning outright, as some people do.

A serious weakness of some public campaigns is that they focus on obvious problems such as polluted air and water, and hypothetical hazards about genetic engineering, and overlook less evident difficulties that may be more important in the long run. Scientists are acutely aware that today's rate of growth of the world population will soon lead to much greater food shortages. It is probable that disaster will only be averted by the use of nuclear and agricultural technologies that are now popularly held to be undesirable. Scientific and social aspects must be balanced.

Population control is another matter of the greatest importance. In the last sixty years the world's population has tripled, from about two billion to six billion. Future technologies (including nuclear technologies) will be unable to provide enough food to feed the world if the population continues to grow without check. New discoveries cannot enable us to cope with a population that increases indefinitely. Are we justified in gambling on that possibility, when widespread human misery and starvation will result if the necessary technical results are not achieved? Surely not; we must control the increase of population.

Population control must be considered in relation to the welfare states of the developed world. By a welfare state I mean one that, at least as a matter of policy, accepts some responsibility for the welfare of its citizens, and at the very least tries to prevent them from starving. Some conservative people seem to dislike the expression "welfare state," thinking (correctly) that it was brought about largely by the efforts of persons of liberal views. (In some quarters "liberal" seems to have become a term of abuse.) But does any reasonable person think that society should return to the policies of the nineteenth century, when countries that considered themselves civilized allowed many of their citizens to starve to death? Some form of welfare state must be our aim for the whole world.

The dilemma we face is that the welfare state on the one hand, and failure to control population on the other, are incompatible; they will inevitably lead to a population explosion. If people are not encouraged to adopt birth-control procedures in welfare states, the result will be worse misery than has ever existed. It is unrealistic to put the blame on individuals who produce more children than they are capable of providing for. Greater blame must lie with powerful organizations, like the Roman Catholic Church and some of the non-Christian churches, which, in the name of a narrow-minded morality, condemn birth-control methods and keep millions of people ignorant of those methods and unable to obtain the necessary facilities. If they will take a broader view they will see that they are violating a higher morality by leading to a society in which there will be suffering of a magnitude never yet encountered.

We all need to have informed opinions on the possible disasters that may befall our Earth. Some of these, at the present state of technology, we can do nothing about. If a massive meteorite is rushing toward Earth, we have no practical way of heading it off; perhaps in later centuries the necessary techniques will become feasible.

We must be concerned about the possibility that what we do today will create difficulties for our descendants. One problem faced by medical scientists is that some modern procedures for treating diseases may inadvertently encourage the onset of new diseases. Scientists have warned that the overuse of certain pharmaceuticals gives rise to more virulent forms of bacteria and viruses.

*   *   *

A more immediate problem is the possibility of global warming, which has undesirable consequences like weather changes and a rise in the level of the oceans. Theories of how this may occur have a long and interesting history. In 1824, French mathematician Joseph Fourier (1768–1830) compared the role of the earth's atmosphere to a bowl covered by glass and heated by the sun, the effect being like in a greenhouse—although the expression "greenhouse effect" seems to have been first suggested much later, in 1937. The principle is easy to understand. If we stand in a greenhouse with the sun shining on us, we can feel directly the heat brought by the rays of the sun; we can feel the radiation inside the greenhouse almost as if we were standing outside. The heat from the sun passes freely through the glass of the greenhouse. During the day, the sun heats the air and plants in the greenhouse. At night, when the temperature outside falls, the heat from the greenhouse cannot easily pass through the glass. The reason is that glass does not transmit well the much-longer wavelengths of the radiation from the greenhouse. Thus, the heat can get in easily during the day, but leaves with difficulty at night.

The possibility that carbon dioxide, CO_2, in the upper atmosphere plays an important role in producing this effect was enthusiastically taken up in 1896 by Swedish chemist Svante August Arrhenius (1859–1927). He showed that carbon dioxide acts like glass, allowing the sun's rays to pass freely through the upper atmosphere, but absorbing the radiation of longer wavelengths that try to escape from the earth. Arrhenius considered in particular the production of CO_2 in volcanic eruptions, and suggested that the ice ages of the past were brought about by decreases in the carbon dioxide content of the atmosphere.

It is now generally agreed that CO_2 and certain other gases emitted as a result of industrial processes are accumulating in the upper atmosphere and having some effect on the earth's climate. Many measurements have been made of carbon dioxide concentrations in the upper atmosphere. The quantities involved are enormous, about 14 billion tons of carbon dioxide being added to the atmosphere every year; the total amount present at the turn of the millennium was 1,400 billion tons. Carbon dioxide is not the only greenhouse gas; another is methane, which is produced naturally in agricultural processes, particularly in the processing of rice.

Global warming is a matter of controversy, and exaggerated opinions have been expressed on both sides. A few experts, and many more people with no technical knowledge of the matter, insist that no global warming is to be expected. Others predict dire effects, and as usual the truth is in between—but where? The best that most of us can do is to ascertain the consensus of experts who seem to take an unbiased view.

There is a simple scientific point that is widely misunderstood. It is often stated that global warming will lead to a rise in the level of the oceans because floating icebergs will melt. Such melting, however, has no effect on the level of the water, as anyone can verify experimentally with ice floating on water in a tumbler. It is a matter of elementary physics that when a lump of ice floats on water, the volume submerged is equal to the volume of the water produced when the whole lump melts, ice occupying more space than the same mass of water. Global warming does indeed raise the level of the oceans, and this is due to two causes. The first is the natural expansion of water. The second is the melting of some of the ice resting on solid earth; this includes glaciers and polar ice that is resting on rock.

An admirable and impartial account of global warming is given by Sir John Maddox in his book, *What Remains to Be Discovered: Mapping the Secrets of the Universe, the Origins of Life, and the Future of the Human Race.* Since Maddox was for many years the editor of the scientific journal *Nature*, he has had every opportunity to become thoroughly familiar with the subject. He emphasizes that the consensus among experts who have carried out computer modeling is that, if industrial activity remains the same as today, there will be an average temperature rise of one degree Celsius between 1990 and 2050. The level of the oceans may, as a result, be half a meter (about twenty inches) higher than it was at the turn of the millennium.

This is not as much as pessimists predict, but it is significant and will have important effects on land masses in some regions. In some places the changes will be beneficial, in others, detrimental. The temperature rise may have a significant effect on agriculture; the yields of crops may sometimes be reduced, a serious matter in view of the expected increase in world population.

These predictions about global warming are based on the assumption that the amount of atmospheric pollution from industrial activity will remain the same. However, with the expected increase in the earth's population, the level of industrial activity may rise, with more pollution and a greater rise in temperature. It is thus important to restrict the pollution level, and fortunately, many nations are actively concerned in this endeavor. I think there should be a change in policy in the use of nuclear energy. Nuclear power plants produce much less pollution than generators that use fossil fuels. Some countries have set a good example by abandoning the old-fashioned methods in favor of nuclear power. Other nations, however, including the United States, Canada, and Britain, have allowed ignorant propagandists to affect their policies, and are building few nuclear power plants. It seems to me that these policies must be reversed if our descendants are not to be faced with a serious global calamity.

*   *   *

Before we leave scientific matters and go on to more general ones, it is worth considering how pure scientific research, which advances basic scientific knowledge, has led to practical consequences. Those who award grants for scientific research place great stress on its possible useful applications, but this is a short-sighted policy. It is quite impossible to make reliable prediction of the practical consequences of a piece of fundamental research. Indeed, even after quite practical devices were first developed, it was often thought initially that they would not be much use; table 3 on pages 158–59 shows examples of this with respect to the telephone, radio, home computers, and other inventions.

Many of the consequences of a scientific discovery could not possibly have been recognized at the time the work was done. Without much thought of any practical applications, Michael Faraday carried out investigations of electromagnetic induction, which made possible the large-scale production and distribution of electrical power. Clerk Maxwell developed a theory of electromagnetic radiation without any thought of radio transmission. Even Heinrich Hertz, who first transmitted electromagnetic radiation (effectively the first radio signal) in 1887, expressed the opinion that one could never broadcast over more than a few meters. Lord Rutherford, whose work directly made possible the utilization of nuclear energy, at first ridiculed the idea of any practical application.

The work of chemist Chaim Weizmann (1874–1952) provides a particularly instructive example. Born in Russia, he first worked in Germany and Switzerland and obtained his doctorate at the University of Fribourg. He became active in the Zionist movement, and moved to Manchester in 1904, partly because he thought that Britain was the most likely country to bring about the establishment of a Jewish national homeland in Palestine. He worked at Manchester University with professor of chemistry William Henry Perkin, Jr. (1860–1929).[2]

Weizmann decided to work on the general problem of the synthesis of organic compounds by bacteria.[3] He had no practical aim in view; he just thought that there should be more basic knowledge on the subject. He worked with a bacterium now called *Clostridium acetobutylium* and discovered that it would convert sugar into butyl alcohol. However, when he told Perkin, the reply was "Butyl alcohol is a futile alcohol; throw it down the sink." Perkin had little sense of humor, but apparently even he could not resist this little joke.

Weizmann, however, found that the bacterium also produced acetone. At the time this was of little use, since acetone could be produced in sufficient quantities from wood. However, when World War I broke out there was an urgent need for acetone to manufacture gun propellants, and not enough could be obtained from wood. In the end, through the intervention of Winston Churchill, a gin factory was taken over, Weizmann was appointed its chief chemist, and large amounts of acetone were made using the bacterium. Some of this work was done in Toronto, where they used corn for the production of the alcohol.

Weizmann's technical success had much to do with the help given by the British government toward the establishment of a Jewish homeland, of which he became the first president in 1948. This incident inspired a three-act play by George Bernard Shaw, written in 1936 and called *Arthur and the Acetone*. "Arthur" refers to Arthur Balfour, the British statesman who was responsible for the famous Balfour Declaration of 1917, which promised the homeland in Palestine. I am not aware that the play has ever been staged.

Weizmann's bacterium had yet another success, because butyl alcohol was very far from futile. It is an excellent, fast-drying paint solvent, and is used extensively in the automobile industry.

It is clearly folly to expect pure scientists to justify their intended work on the basis of utility. Almost invariably, practical consequences can be identified only after, not before, the research is done. The sole criterion for approving a scientific project should be its quality.

*   *   *

We all recognize that it is essential, when dealing with the problems of life, to make sure that our emotions do not override our intellects. Emotion should not be suppressed, but disciplined and controlled, so that we are its master and not its slave. Unhappy human relationships, such as the short-lived marriages that are so common at the beginning of the twenty-first century, are often due to a failure to keep passions under the control of reason.

The world has become so extraordinarily complex that many factors have to be considered when making decisions. From time to time, we all have to make decisions about practical matters—for whom to vote, whether to attend a university, whom (if anyone) to marry. Some are of far-reaching importance. A wrong decision about whether to pursue higher education may make all the difference between a happy and an unhappy life. There is an enormous amount of luck in all our lives, and the prudent person tries to navigate a successful path by making sensible decisions. We must realize—as expressed so well in the quotation from T. S. Eliot at the beginning of chapter 1—that information is not knowledge or wisdom, but requires processing by our intellects.

I would like to mention here some of the obvious pitfalls that we encounter in dealing with information, scientific and otherwise. The most important of these lies in the use of words. We do most, if not all, of our thinking with words. We gain our information from the spoken or written word, and we cannot avoid using words in processing it. When we think we have reached a conclusion, it is always helpful to formulate it in writing. As Francis Bacon pointed out in his essay, "On Studies," "Reading maketh a full man, conference a ready man, and writing an exact man."[4] The great physicist Ernest Rutherford often pointed out to his students that a piece of research was not completed until it had been written down in clear English.

There are serious pitfalls in the use of words; in the first place, words often have several meanings. A good example is provided by the word "love," for which the *New Shorter Oxford Dictionary* gives ten distinctly different meanings. The spectrum of meanings is broad, including nothing (as in tennis) and several quite different forms of emotion. A few words have exactly opposite meanings; the word "cleave," for example, means both to split apart and to stick together. Most words, including many of the simplest ones, have multiple meanings, so that when we are doing any serious reading, we should from time to time consider whether we are understanding words in the sense that the writer intended.

Also, words often mean different things to different individuals and in different contexts, and this is easily exploited by unscrupulous people to give an entirely wrong impression. A delightful example of word-play appears in a passage from Charles Dickens's *Pickwick Papers*, in which a judge had elicited from Sam Weller the admission that it was probably Sam's father who had created a disturbance in court:

> "Do you see him here now?" said the judge.
>   "No, I don't, my lord," replied Sam, staring right up into the lantern
> in the roof of the court.

Sam told the strict truth, but completely misled the judge, who had not been careful in wording his question.

One of the most common ways of misleading people with words is in the selection of information. As Macaulay pointed out in chapter 1 (see page 21), it is easy to string together bits of correct information and point to an entirely false conclusion. We can select correct information so they arrive at an apparently convincing argument against the use of nuclear energy, and then can take another set of correct data to support the contrary argument.

Another way to mislead by words is to make statements that are devoid of meaning but that, *because of their style*, delude readers into thinking that something profound has been said. This trick is often used by people who should know better, such as scientists and scholars—although not by the best of them. Theologians and philosophers are particularly apt to play tricks with words, sometimes deliberately, or perhaps more often because they have become used to doing so because of practices that are common in their fields of study. The so-called scholastic philosophers who began to flourish in the twelfth century, and whose influence is still felt, were particular addicted to arguments that, on careful analysis, become meaningless. English philosopher John Locke (1632–1704) was one of the first to call attention to the shallowness of their arguments, particularly in his *An Essay on Human Understanding* (1690).

The debates of these philosophers, he observed, passed "commonly for Wit and Learning, but were really a cheat." He wrote at length and amusingly about "this artificial Ignorance and *Learned Gibberish*, prevailing mightily in these last Ages." It is disappointing to find that these confusing verbal practices are still prevalent today.

As I mentioned in chapter 1, with a choice example, a serious offender was the French paleontologist Teilhard de Chardin (1881–1955), who had many admirers in his time, chiefly among intellectuals who did not know much about science, but were apparently bemused by Teilhard's unintelligibility. In a devastating review of his book, *The Phenomenon of Man* (1955), the British immunologist and Nobel laureate Sir Peter Medawar (1915–1987) wrote:

It is written in all but totally unintelligible style, and this is construed as prima-facie evidence of profundity. . . . It is because Teilhard has such wonderfully *deep* thoughts that he's so difficult to follow—really it's beyond my poor brain but doesn't that just show how profound and important it must be?"[5]

It is to be regretted that, even recently, a few scientists have published books that seem remarkably obscure to many scientists, but have been popular with the general public.

Examples of verbal obscurity are lavishly provided by a group referred to as *postmodern positivists*, whatever that may mean. The fad apparently began in France, but has spread to other countries, and is particularly widespread in universities in the United States. There are also people called *cultural relativists*, and since they are equally unintelligible, I do not know the difference between the two groups. Since the forte of all these people is obscure writing, it is difficult to know whether their philosophy has any basis. They take delight in disparaging science and its methods, but, paradoxically, are particularly fond of introducing scientific language into topics to which it cannot apply.

What should have been the death blow to these fads was administered in 1996 by Alan Sokal, professor of physics at New York University. He wrote an article entitled "Transgressing the boundaries toward a transformative hermeneutics of quantum gravity," which parodied the style of postmodern relativists. Any knowledgeable person with a little scientific knowledge can see at once that the paper is full of scientific nonsense, written in a pompous and fatuous style. Sokal submitted the paper to the journal *Social Text*, and it was accepted and published; the editors were completely taken in. One would have expected that this incident would have discredited the whole field of postmodern positivism, but wrongs do not right themselves so easily.

Sokal's paper is reproduced in full in a book he wrote with the French theoretical physicist Jean Bricmont; it was first published in French, the British version appearing with the title *Intellectual Impostures*, and the American one with the title *Fashionable Nonsense*.[6] The book is a brilliant exposé of the fad, with many superb examples of the kind of gibberish it generates.

Obscurity of expression usually indicates muddled thinking, but many people do not realize this. During my long academic career, many students have commented to me that some professor gave very confused lectures, and have then added that, of course, he or she must be very brilliant. How kind to take such a charitable view of bad lecturing, when in almost every case the lecturer was really not at all competent.

It is true that a few great thinkers, like physicists Clerk Maxwell and Neils Bohr, have been bad lecturers. In 1939 I attended a lecture by Bohr at

Princeton University, and it was by far the worst lecture I have ever heard. I sat in the front row, but he spoke so quietly that I could not hear him; his lettering on the blackboard was so small that no one could read it. It is hard to imagine a worse lecture.[7] My experience in general, however, is that distinguished scholars and scientists usually express themselves clearly. My advice is that, rather than admiring a person for writing or lecturing obscurely, we should bear in mind the strong possibility that they are incompetent.

Words are tremendously important, essential to communication; even pure mathematicians have to use words. It is essential to use words as correctly as possible, putting them together according to the rules of grammar and syntax. There is a popular fad, perpetrated and sometimes initiated even by professors of linguistics and literature, which proclaims that people should be allowed to speak and write in any way they wish. This can only lead to confusion. People who speak and write carelessly will never clarify their own thoughts and will inevitably be misunderstood. Since thinking is done with words, it follows that unless we use words grammatically we will not think straight.

At the same time, we must not become slaves to words. Many apparently irreconcilable arguments between people are futile and fruitless since they involve no more than definitions of words. Many philosophical, ethical, and theological disagreements are like this. The question of when an embryo becomes a "person" depends solely and critically on how one defines a person. That seems obvious, but it is surprising how many people will argue about the matter without ever facing the problem of definitions. Words like "belief," "existence," "truth," and "God" are particularly treacherous. More will be said in the last chapter about the "truth" of science.

A prevalent confusion is between concrete and abstract nouns. A concrete noun refers to something that can be observed and measured, such as *book, elephant,* and *atom.* Abstract nouns refer to unobservable notions, such as *difficulty, idea,* and *leadership.* The distinction seems straightforward, but since words often have different meanings, some nouns can refer in different contexts to concrete objects or to abstract ideas; examples are *structure* and *music.* This often leads to confusion.

If a word can only be an abstract noun it is important not to use it as a concrete one. A good example is provided by the word *leadership,* which, as the suffix *–ship* tells us, is always an abstract noun. Often, however, we hear someone say that the "leadership decided" something, when what was meant was that the "leaders decided." This example is perhaps innocuous, since it evokes irritation rather than confusion, but it is bad to get into the habit of misusing concrete and abstract words. Abstract words like *rights, ethics,* and *patriotism* are often used loosely today, often to clinch an argument without any real evidence. Telling people that to argue against war shows a lack of patriotism is too often used by leaders to wage wars that lack justification. Dr. Samuel Johnson commented: "Patriotism is the last refuge of a scoundrel."[8]

When parties are locked in argument, they should first make sure they are using words properly, for that is the commonest form of misunderstanding and often easy to resolve. Until it is resolved, the contending parties are only shadowboxing. Once it is settled, it is possible to see what the differences of opinion really are.

*   *   *

Other items of dispute are ideological, and they have many causes. One of them has been called the *argument from personal incredulity* by Richard Dawkins. In *The Blind Watchmaker*, Dawkins criticizes the objections raised by a bishop to some aspects of the theory of evolution.[9] The bishop's argument was peppered with phrases like "it is impossible to believe," "it is hard to understand," and "it is not easy to see." Such phrases often indicate that the writer's objections may well be based on personal ignorance of the field of study. In one example, Dawkins shows that this was indeed the case. The bishop found it difficult to understand how the whiteness of polar bears could be explained in terms of natural selection, since polar bears have no natural predators. The answer (as any university student of biology would know) is that polar bears hunt for seals on the ice, and do so more effectively if they are white.

The argument from personal incredulity was also used by philosophers at Oxford who criticized Einstein for his theory of relativity; they doubted its validity because they could not understand it, but they were not well-versed in physics and mathematics. They also failed to realize that Einstein's theory of relativity has nothing to do with the philosophers' concept of relativism, a question of definitions again.

The motto of the Royal Society of London is *Nullius in Verba*, which, roughly translated, is "Take nobody's word for anything." The motto is perhaps a short version of Horace's "*Nullius addictus iurare in verba magistri*," which means "not bound to swear by the word of a master." This is wise advice; as far as possible we should make up our own minds and not put blind trust in individual authorities, who have often been wrong. Of course, if we are trying to form an opinion about a science in which we have no competence, we are obliged to follow the advice of disinterested experts. I personally have no special expertise in biology, but when I find that all competent biologists accept the theory of evolution, that some of them, like Steve Jones and Richard Dawkins, can write clearly and logically about it, and that the theory has led to important technological advances,

I am satisfied that it must be essentially correct. On the other hand, I see no reason to accept the authority of any religious or philosophical writers about evolution. We may sometimes find their ideas helpful, but we should also examine them critically.

By the time we reach our twenties, all of us have adopted a certain attitude of mind and a style of living that, sometimes, remain with us to the end of our days. This style includes our use of language, our taste in music and other forms of art, and the kinds of things we like to do in work and play.

The type of individual we become depends on both our genetic inheritance and our environment, particularly in our early years. It was not a scientist, but William Shakespeare, who seems first to have called attention to nature and nurture. In *The Tempest* he has Prospero say of his adopted son, the savage Caliban:

> A devil, a born devil, on whose nature
> Nurture can never stick; on whom my pains,
> Humanely taken, all, all lost, quite lost.

"On whose nature Nurture can never stick" is a splendid phrase, full of meaning. Today it may conjure up the idea of software being loaded onto hardware in a computer. Computers indeed offer a useful analogy to the nature-nurture situation. Our genes act somewhat like a computer's hardware, which is made up of the mechanical parts that we can see and touch if we open the box. Our nurture acts like the computer's software, which comprises the programs, operating information, and data that we feed into it. If anyone were to ask us to state the relative importance of the hardware and software, we would not be able to give a useful answer; indeed, we would think the question rather naive. After all, we might have superb hardware in our computer—"cutting-edge" in the modern jargon—but if we have inadequate software the results will be poor. If we load excellent software into a computer with bad hardware, the results will again be poor. For the computer to operate well we must have at least adequate hardware *and* software. The situation with nature and nurture is analogous. For a human being to lead a productive life, both nature and nurture must be satisfactory. A person born with highly satisfactory genes (good hardware) may, if badly brought up (bad software), develop into a criminal. A person with genes tending toward criminality might, with a suitable upbringing, turn out to be a model citizen; the criminal tendencies may have been diverted into highly moral activities.

It is a fallacy to try to quantify nature or nurture, or say that one is more important than the other. It is only legitimate to quantify if nature plays the only role and nurture plays no part. This is true with a number of genetic disorders. Color blindness, for example, is entirely determined by genes. But if both nature and nurture are involved, it makes little sense to speak of their relative importance. Even some experts fail to understand this point, and think they can make estimates of the relative importance of the two factors.

The fundamental reason why it is impossible to quantify genetic and cultural influences is because they interact, just like the hardware and software in our computers. So they are not *additive* quantities. Simple additivity applies to numbers of objects. If we have two apples and obtain three more, we have five apples. If we have two apples and three pears we can say that 40 percent of our fruit is apples. But consider the color of something on a screen, or of an object. Anyone who has worked with paints knows that the colors are not obtained additively from colors that are mixed together.

The point is that if two effects *interact* with each other they are not additive. This is true of the hardware and software in computers. It is also true of heredity and environment. We cannot quantify their relative importance, since they interact with each other in a complex manner that we still don't understand. Some of the physical and mental characteristics we are born with, such as eye color, color blindness, and some mental disorders, are influenced entirely by genetic factors. Other physical qualities and mental attributes that we develop in later life are partly genetic and partly environmental. Some people, for example, are born with a predisposition to develop lung cancer, and under some environmental conditions (for example, if they smoke) they do develop the cancer. Other people may smoke just as much and live under worse environmental conditions, but never develop cancer; their genes did not give them a tendency to develop the disease. Obviously, we cannot say that the effect is so many percent genetic and the rest environmental; the interactions are complex and not additive.

The situation is most straightforward (but still complicated) with certain medical disorders that can be related to single genes, thus illustrating the OGOD principle: *one gene, one disease*. This is true for over one thousand physical and psychological disorders, like color blindness, cystic fibrosis, hemophilia, and Huntington's disease. These genetic identifications simplify matters for medical practitioners, who can sometimes diagnose the disease by a biochemical test. Perhaps cures can someday be achieved by gene replacement.

Often, however, disorders are caused by a number of genes acting in a concerted manner. In mice, the texture of hair and skin has been related to a group of seventy-two genes, and no doubt the same is true of humans. Cancer has been related to more than a hundred genes, and it is expected that many more will be found. Shyness has been related to two dozen genes. Environmental factors play a role in these conditions, but we cannot quantify them; nurture and nature are not additive.

We are all born with certain predispositions, and how we lead our lives depends on these and also on the influence of our culture. For example, some people are born with strongly aggressive attitudes. If they are brought up in a benign environment that does not encourage aggressive behavior, they may be able to control their aggressions and behave in a reasonable way. If, on the other hand, they are encouraged by their environment to behave aggressively, there is little chance that they will not be aggressive. Conversely, a person born with little predisposition to aggressiveness may remain unscathed even if brought up in an atmosphere of aggression. At the present time, the great amount of violence in the world presents a serious problem. Some of it is due to the proliferation of weapons. Some of it is due to changes in culture; young people who have seen many murders on television have more of a tendency to be violent themselves. A child born with aggressive and intolerant proclivities is more likely, by the principles of heredity, to have parents who are prone to violence, and thus grows up in a violent subculture. In other words, often there is positive feedback as far as nature and nurture are concerned.

Toward the end of chapter 3 we discussed positive feedback, giving as a familiar example the sound level at a cocktail party. Since positive feedback is found with aggressive and intolerant attitudes, it leads to many of the ills of today's society, such as high crime rates, the sectarian violence in Northern Ireland, and incidents in the Middle East. If the world is to become a happier place, our first goal should be to encourage tolerance and civil behavior.

Positive feedback plays a powerful role in another way: as discussed earlier, it sometimes leads to what scientists now call *chaos*, not utter chaos but noncatastrophic chaos. The essential characteristic of this kind of chaos is that the outcome is unpredictable, however well we know the initial conditions. Unpredictable behavior is very common in everyday life. A war is a good example: who could have predicted that the assassination in 1914 of the Austrian Archduke Franz Ferdinand by a Bosnian student would precipitate World War I (if indeed it did so), and who could have predicted in any detail the course and outcome of that war? Did the length of Cleopatra's nose really influence the course of history, as Blaise Pascal suggested? Would World War II have occurred if Adolf Hitler had been run over by a bus when he was a house painter? Will the instability of the world be increased or decreased if some of the Western nations attack a Middle Eastern country?

What should be done to make our rather sad world a better place? The analogy of the computer's hardware and software may help us here. If a computer is unsatisfactory, we take it to an expert, who may suggest modifying the hardware, changing the software, or both. The first is analogous to modifying a person's genes by genetic engineering, the second to changing the environmental effects. Knowledge of genetics is still imperfect, so perhaps we are not ready to meddle with the genes of persons of limited intelligence or those convicted of criminal offenses. In the future it may be considered safe and humane to do such things, but today we must concentrate on modifying the environment as far as possible.

Luck seems to play an enormous role in our lives. Many successful people have said that a chance meeting with someone, or a chance remark, caused them to choose their careers; without it they might well have done something quite different. The influence of family background is particularly strong. Two boys may be born with similar genetic predispositions, both with a strong desire to become successful in life. One may be brought up in a musical family, and as a result becomes a distinguished composer. The other, with a gambling gangster for a father, may be led to a life of crime. The genetic differences between the two boys may have been small, but their effects have been magnified by their different environments and by positive feedback.

Cultures and subcultures, like genetic characteristics, are transmitted from generation to generation. It might seem that this is an example of the inheritance of acquired characteristics, suggested by Lamarck but now believed to be unimportant. There is no suggestion, however, that culture is passed on by genes. Instead, the transmission of culture is an environmental effect. As time goes on, the behavior of groups of people becomes modified, and individuals are correspondingly affected.

Whereas genetic characteristics become modified through the generations at a snail's pace, the transmission of culture occurs much more rapidly. The English language of today is quite different from that used by Geoffrey Chaucer in the fourteenth century, but human beings are much the same. Chaucer and a modern English-speaking person would find it exceedingly difficult to communicate with each other—they would first have to study each other's use of language. This may seem surprising but consider how easily a message becomes distorted if it is passed from mouth to mouth. There is an old story from the First World War. The message "Send reinforcements, we're going to advance," when passed along a chain of soldiers, ended up as "Send three and fourpence, we're going to a dance."

In *The Selfish Gene*, Oxford zoologist Richard Dawkins introduced the useful concept of the *meme*, which is to culture what the gene is to heredity.[10] Just as the genes are the replicators that bring about the transmission of characteristics from one generation to another, the memes are concerned with the transmission of culture; they are units of imitation. This is no more than a useful analogy, since there is as yet no evidence that the memes have the same physical significance as genes; they do not seem to be identifiable as chemical structures, whereas the genes are portions of DNA molecules and therefore of chromosomes.[11] It is useful to imagine that there are various specific memes that replicate in much the same way as genes, but that lead to a much faster evolution of a culture. We can suppose that there are memes for fashions in clothes, for music, and for religion. It is far from obvious what a single meme consists of. There can hardly be a single meme for music, but rather a large group of unit memes; perhaps musical creativity, a sense of pitch, and other musical skills might each be regarded as related to a meme. All of them are also related to corresponding genes whose effects have been amplified by interaction with the cultural environment.

Regarding the improvement of the human race, some memes are good and others bad. The memes associated with racial and religious tolerance are, from that point of view, good. The memes leading to the attitudes of terrorists, with their desire to exterminate people of a different race or religion, are obviously bad. Unhappily, the corresponding memes are likely to persist for many generations because of positive feedback; children brought up in an atmosphere of intense racial or religious intolerance are unlikely to change their ideas. We may consider a bad meme to be a kind of virus that invades the mind. Sometimes the mind, because of genetic influences or the effects of other memes, develops a degree of immunity that renders the bad meme powerless. In other cases, the influence of the bad meme is overwhelming.

Although there are many uncertainties about terrorism and other criminal actions, there is strong evidence that racial factors do not play a predominant role, as is sometimes suggested. The unfortunate events in Northern Ireland seem to point to that conclusion. The people of Ulster are racially similar to those in the United Kingdom and North America, where Roman Catholics and Protestants are able to live in harmony. But in Ulster, Catholics and Protestants have been shooting one another for centuries. The cause can hardly be one of race, since they are all of much the same race. Cultural differences, resulting from their religious connections and reinforced by positive feedback, must be a reason. The unrest in the Middle East is also obviously greatly enhanced by feedback.

One rather positive thing about memes is that we all have the possibility of passing on our memes to later generations, although few of us do. This is not true of our genes, since at each generation the contribution of our genes is halved; we cannot control that in any way. Prince Charles is a linear descendant, by thirty-one steps, of William the Conqueror, and has exactly the same genes (as do all human males), but has completely different alleles (varieties of genes).[12] Since the contribution of genes is halved in every step, it follows that, in thirty-one steps, the fraction remaining is one divided by about two billion. Since we have about 40,000 genes, this means that it is highly unlikely that any of William's genes will have been passed on to Prince Charles, even allowing for the fact that he is descended from William by many paths. Similarly, if Aristotle, Socrates, or Shakespeare have any living descendants, it is highly unlikely that any of their genes will be an exact match. Their memes, however, are still playing a memorable role in our culture.

*   *   *

The way culture develops in a society is complex, and I can do no more than comment on a few points. The factor now having the strongest influence on culture is the advance of technology. We see a striking example of this in the way people spend their leisure hours today as compared with a couple of centuries ago. Now we are able to see televised dramatizations of novels written in the early years of the nineteenth century, including Jane Austen's novels. One cannot fail to be struck by the fact that in those days people had to devote much more effort to creating diversions for themselves by simple means; an evening's entertainment might consist of dancing to the music provided by a piano and perhaps a few other musical instruments. It was not only the "gentry" who behaved that way; poorer people enjoyed country dances and other simple entertainments.

It is difficult to imagine many people today being satisfied by such entertainment, and changes of attitude were no doubt influenced by technological advances. An important role is played by public-address systems, which amplify music and bring it to thousands of people in a single stadium, and, by means of radio and television, to millions of people throughout the world. In Jane Austen's day a musical performance could be heard only by a small group of people—some hundreds at the most in an opera house—because of the lack of sound-amplifying systems. This in itself prevented the rise of the "pop" subculture; many (including myself) would say that people were lucky in that regard.

Radio and television exert a powerful influence on our culture. Radio transmission was possible for at least two decades before it occurred to anyone that it might be used for entertainment. At first its sole use was to supplement the telegraph, its main advantage being that it could communicate between ship and shore. The telephone has always been used almost exclusively for communication. The only early exception I know of is that an 1880 choral concert in Zürich was transmitted about eighty kilometers (fifty-five miles) to the telephone office in Bâle, where a large audience is reported to have heard it "with great fidelity."[13]

When it was suggested in the early 1920s that radio programs might be received by the public, there was general derision. Asked at that time to invest in such a project, David Sarnoff, the president of the Radio Corporation of America (RCA), is reported to have said, "The wireless music box has no imaginable commercial value. Who would pay for a message sent to nobody in particular?" Soon afterwards, however, the broadcasting of news bulletins, lectures, and music became popular. At first, the contents of transmissions were, for the most part, under tight government control. The British Broadcasting Corporation regarded itself as the guardian of morals and culture. The newsreaders wore evening dress (even for radio), and the public was allowed to hear only good grammar (certainly no blasphemies or obscenities) and highbrow music. Much the same was true when television first became popular. In an article in 1933, the radio pioneer Sir Andrew Fleming expressed the official attitude to this new technique: "Let us hope that this new weapon science has provided will not be vulgarized or put to base use, but enjoyed for the instruction, elevation, and national entertainment of the public at large."

Today all that is changed. Much of radio and television has become commercial, and programs are partly controlled by the will of the people, and perhaps even more by the preference of commercial sponsors. There is obviously a great incentive to broadcast programs that are popular, regardless of their quality. In 1969 Clive Barnes, the well-known theater critic of the *New York Times*, made a significant comment: "Television is the first truly democratic culture—the first culture available to everyone and entirely governed by what the people want. The most terrifying thing is what the people do want." It is still possible to receive some radio and television programs of high quality, but the vast majority are what I personally regard as unappealing. Of course, I am only expressing the views of a person with a particular set of genes who has lived in a particular environment.

We can sum up by saying, that although we believe that we can exert free will, we are all greatly constrained by our genes and our memes. It seems that some people become complete slaves to their genes and memes. Macbeth thought so when he said:

Now I am cabin'd, cribb'd, confin'd, bound in
By saucy doubts and fears.

We cannot yet do much about controlling genes, and the control of culture is of limited effectiveness. All most of us can hope to do is try to make the world a better place to live in.

Chapter 7
Religious Belief

Somebody asked Bertrand Russell [an avowed agnostic] at some meeting "Lord Russell, what will you say when you die and are brought face to face with your Maker?" He replied without hesitation: "'God,' I shall say, 'God, why did you make the evidence for your existence so insufficient?'"

A. J. Ayer, *The Standard* [a British newspaper], 1984

Reaching a satisfying conclusion about religious belief is, for an increasing number of us, the most difficult task of all. Religious belief is not something that we can acquire by the use of reason. We cannot simply consider the world around us, select wisely from the vast amount of information available, and come to a firm religious conviction. Instead, religion is more a matter of personal philosophy and of taste; religious faith is something like a preference for particular kinds of art and music. Although adherents insist that their religions can be given some rational justification, it is impossible to arrive at a religious conviction by rational arguments. As the great Scottish philosopher David Hume put it, religion is founded "on faith, not on reason."

Scientists and others have sometimes said rather harsh things about faith. In *The Selfish Gene*, for example, Richard Dawkins defines faith as "blind trust, in the absence of evidence, even in the teeth of evidence."[1] I would agree with him if he had been defining "blind faith," which perhaps is what he meant to do. I prefer the first two definitions of faith in the *Shorter Oxford Dictionary*: "1. complete trust or confidence. 2. firm belief, esp. without logical proof." We all have certain kinds of faith, or our lives would be bleak. Two people who enter into a prudent marriage have faith in each other based on what they know of each other's characters. Most of us have faith in friends and colleagues: our lives would be intolerable if we did not, and mistrustful people are usually miserable people. I see nothing wrong with faith, as long as it is based on some kind of evidence and is not unreasonable. What is unfortunate is that so many religious people do not seem to have reasonable convictions; their irrational ideas tend to bring religion into disrepute.

Many of us who are not religious in the conventional sense have no difficulty in entertaining the idea of a creator who originally set up the universe and formulated the universal laws of nature. The difficulty that more and more thoughtful people face today is in accepting the idea of a living god who watches over us, responds to our prayers, rewards us for praising him, and judges us to determine our fitness for heaven.

It is also hard for many people to tolerate the sharp differences of opinion, sometimes amounting to antagonism and even leading to bloodshed, between people holding different religious beliefs. These antagonisms are only too apparent in many parts of the world. Equally unacceptable to many of us are the extreme opinions, such as fundamentalism or creationism, that are held by many religious people. The tendency of some religious people to claim that they, and only they, have reached the absolute truth, and that all other religions are false, also tends to discredit religious belief. Again, the practice by some religious groups of "brainwashing" young and impressionable people into accepting their beliefs brings religion into disrepute. Nor do many aspects of the history of religions make edifying reading; the Crusades inflicted many cruelties, which are fully acknowledged by some modern Christians.

In discussing the basic question of belief in God, I think it important not to be sidetracked by matters like this. The position I take is that the undesirable aspects of religion are aberrations, mostly falling into two main classes. One is fundamentalism, the insistence that certain religious writings, such as the Old Testament and the Koran, must be taken literally. The other is the assertion that, of all the religious sects that exist (over 2,000 of them in North America alone), one's own religion is the only true one and everyone should be converted to it. It is this second aberration, of absolute certainty, that has led to so much strife in the world. Some religious people spend much effort in trying to convert nonbelievers, but in my opinion this is a waste of time. Many nonbelievers are tolerant people who are perfectly content to let people practice their religions, and are much less prone to resort to violence than are some who claim to be religious. The real enemy of true religion is the enemy within, composed of unreasonable, intolerant, and aggressive religious people. They are the ones who bring religion into disrepute.

Some people think that science and religion must inevitably conflict, but I think this applies only to the aberrations, namely religious fundamentalism and intolerance arising from a conviction of absolute certainty.

Some religious people feel that one vital component of a religious faith is a cosmology—an account of the origin of the universe, often included in religious writings such as the Old Testament. In my view, however, cosmology is a science, and religion can make no contribution to it. Religious fundamentalists' idea that scriptural writings must be taken literally was discredited by scholars more than a millennium and a half ago. St. Augustine of Hippo (354–430) was particularly concerned about this problem, and pointed out that one could not expect people to be converted to Christianity if they were told things they knew to be wrong. In this age of so-called enlightenment, it is rather depressing to find that ideas that St. Augustine described as "nonsense" so many years ago are still entertained.

Many mainstream churches today reject the idea that early religious writings are to be taken literally, and recognize that they simply represent the ideas current at the time they were written. This is the sensible conclusion, because rejecting scientific arguments about the age of the earth and other matters can no longer be justified; there is too much evidence. A particularly bizarre kind of argument was first expressed explicitly by Philip Henry Gosse (1810–1888), a respected nineteenth-century biologist who, as a member of the austere Plymouth Brethren sect, was a Christian fundamentalist. He was deeply troubled by obvious contradictions between the Old Testament account and scientific evidence, particularly the fossil record. In 1884 he advanced the idea that the Old Testament account was literally correct and that God had deliberately falsified the scientific evidence to test people's faith. The book was treated with great derision by both believers and nonbelievers.[2]

Gosse's ideas were surprisingly supported recently by none other than the Abbé Georges Lemaître, the eminent cosmologist who was the main originator of the big bang theory. In a conversation with several physicists shortly before the Second World War, he insisted that the account of creation in the Old Testament was literally true, and that God, in creating the earth in about 4000 BCE, deliberately introduced radioactive substances, fossils, and other evidence to tempt scientists into deducing a much earlier age, thus testing their faith. When asked why he was so interested in carrying out his scientific investigations, which led to the much greater ages, he replied "Just to convince myself that God did not make a single mistake."[3]

Most thinking people, believers and nonbelievers alike, find such opinions somewhat ludicrous. But at least they are a trifle more reasonable than simply dismissing scientific conclusions in favor of biblical accounts. If treated properly, religion and science need not come into conflict at all; they are concerned with entirely different realms of thought. Science deals entirely with the physical and biological worlds, and religion with the spiritual world—with the relationship between human beings and God. This point of view was clearly expressed in 1860 by the Reverend Frederick Temple (1821–1902):

> The student of science now feels himself bound by the interests of truth, and can admit no other obligation. And if he be a religious man, he believes that both books, the book of nature and the Book of Revelation, alike come from God, and that he has no more right to refuse what he finds in one than what he finds in the other.[4]

What I like about this statement is that, though made by a churchman who later became Archbishop of Canterbury, it suggests no criticism of persons who are not religious. It recognizes that religious belief is something that some of us have and others do not, and that either position is acceptable. It also makes it clear that one's religious belief should be quite independent of one's judgment on scientific matters.

This view was expressed by Frederick Temple in St. Mary's Church, Oxford, on 1 July 1860, the day after the famous confrontation over evolution between biologist Thomas Henry Huxley (1825–1895) and the Bishop of Oxford, the Right Reverend Samuel ("Soapy Sam") Wilberforce (1805–1873).[5] It is worth discussing this debate briefly, since it led to a serious, long-standing rift between religious people and nonbelievers. Wilberforce based his objections to evolution largely on the basis of scientific arguments rather than religious ones. Both protagonists in the debate were somewhat at fault; Wilberforce based his objections on bad science, and Huxley made an intemperate attack on religion.

The debate occurred about a year after Darwin published *The Origin of Species by Means of Natural Selection*, the first public announcement of the theory. He knew that some religious people, especially those inclined toward fundamentalism, would think that his theory was in conflict with religious belief. He was therefore greatly encouraged by a letter he received from the Reverend Charles Kingsley (1819–1875), who is now remembered as the author of *Westward Ho!* (1855) and *The Water Babies* (1863). Kingsley admitted that the theory had required him to modify his religious ideas, but added that he found it "just as noble a conception of Deity, to believe that He created primal forms capable of self development . . . as to believe that He required a fresh act of intervention to supply the *lacunas* which He Himself had made."[6] Darwin was delighted at this interpretation, and included it in the second edition of his book, which had to be issued at once, because the first edition quickly sold out.

The confrontation between Bishop Wilberforce and Huxley almost never happened, because Huxley had decided to leave the meeting earlier but changed his mind (Darwin was not present; he was habitually indisposed and, in any case, disliked meetings). The debate occurred in the Oxford Museum before an audience of more than seven hundred, on the evening of Saturday, 30 June 1860. Wilberforce was a rather remarkable man; he had taken honors in mathematics as well as classics at Oxford, and had some amateur knowledge of science. He was an eloquent and forceful speaker, and threw scorn on what he regarded as an assertion of the theory of evolution, that humans are descended from apes. During the course of his comments he turned to Huxley and said, "I should like to ask Professor Huxley . . . as to his belief in being descended from an ape. Is it on his grandfather's or his grandmother's side that the ape ancestry comes in?" According to accounts, the Bishop's remarks, which took half an hour or so, were greeted with much applause and cheering.

Unfortunately, he had made a serious error of judgment in choosing such a formidable antagonist. Huxley, aged thirty-five at the time of the meeting, already had a great reputation; for six years he had been professor of natural history at the Royal School of Mines in London. He was far from amused by Wilberforce's question, which was partly said in jest; he was angered by it, and he was an outspoken atheist. Before rising to reply to Wilberforce, Huxley muttered to his neighbor, "The Lord hath delivered him into mine hands," ironically choosing a scriptural metaphor. He forcefully demolished Wilberforce's objections, exposing his inadequate understanding of the subject. His irreverent reply to the Bishop included the passage, "I should feel it no shame to have risen from such an origin. But I should feel it a shame to have sprung from one who prostituted the gifts of culture and eloquence to the service of prejudice and of falsehood."

It has always seemed to me unfortunate that this rather acrimonious confrontation ever occurred, since it led to so much ill feeling between religious believers and nonbelievers. Now that the dust has had plenty of time to settle, thoughtful and tolerant people on both sides of the debate find it possible to live in harmony, and it is to be hoped that this attitude spreads. A comment by Albert Einstein is interesting. After his theory of relativity appeared, the Archbishop of Canterbury at the time, Randall Davidson, was told that he should pay attention to the theory, since it would affect his religious belief. The Archbishop took this advice seriously, but found that he could not understand the theory at all. Later he happened to be sitting at dinner next to Einstein, and discussed the matter with him. He was relieved when Einstein said, "Do not worry; my theory is a scientific theory, and as such has nothing whatever to do with religion."

That is the view of most thoughtful people today, whether they be religious believers or nonbelievers. No conflict need exist between good science and sound religion. Religious bodies should accept sound scientific evidence and conclusions, and concentrate on matters that relate to religion.

*   *   *

As a scientist, I am convinced that the only effective method for investigating the physical and biological aspects of the world around us is by strict observation and experiment. A few scientists believe that there is no more to life than what can be learned from science, but I am not of their number. I believe that there is a spiritual aspect to our lives as well as a physical one. I do not, however, believe that supernatural events have ever occurred, since there is no credible evidence for them. I think it likely that the laws of nature have been obeyed at all times. The establishment of these laws of nature, and of the initial conditions that made our universe possible, may have been brought about by a divine creator, but he seems to me to have left no evidence of his existence. Although I have studied science both intensively and extensively, I can see no "evidence of design." [7] I can find no reason to believe that there is life after death, and consider it a waste of time to pray or to belong to any religious body. I would describe myself as an agnostic with a strong tendency toward atheism. [8] In other words, I keep an open mind, but consider the concept of a god who watches over us and judges us to be beyond belief, because I can find no evidence for it.

While maintaining this point of view, I respect the moral codes of some religions, and the disciplined way of living to which they sometimes lead. At the same time, I have found that many nonbelievers have moral codes that are just as estimable. In fact, I have frequently been struck by the fact that so many nonbelievers are working toward peaceful solutions of the world problems, while so many who profess to be religious favor hostility toward other nations.

I greatly respect the many people who practice their religions in a reasonable way. What I have found surprising is that some of them agree with my position almost completely. Many religious people agree that no supernatural events have ever occurred; some believe that a few have, but consider that, on the whole, the universe unfolds in accordance with the laws of nature. They nevertheless find that practicing their religion fulfils an important spiritual need. In other words, the difference between us is one of personal taste rather than of reason; they have faith, and I have not.

On reflecting on the universe around us, I find myself unable to accept the hypothesis that God intervenes in the unfolding of the universe and in our lives. In other words, it seems to me that if there is a god, he did no more than set up the universe. Consider the earth's population, and the condition of the people of this world. At the beginning of the third millennium the world's population has reached about six billion. About three billion people—half of the world's population—live in abject poverty, with inadequate water, food, sanitation, and medical care, and without any hope of anything better. Of these, about one billion suffer from severe malnutrition. About 80 percent of the people on this earth live in what are classified as underdeveloped countries, which means their lives are far less than comfortable. Every year, nearly twenty million people, mainly children, die of starvation; this number is greater than the population of, say, Sweden. Dozens of wars rage over the surface of the earth, terrorism is rampant, and many millions live in daily terror for their safety and their lives. I have to agree with what Tom Stoppard wrote in *Rosencrantz and Guildenstern Are Dead* (1970): "Life is a gamble at terrible odds—if it were a bet, I wouldn't take it."

These facts are more than enough to convince me that if God exists, he plays no active role in the world today. Another circumstance leads me even more forcibly to this conclusion. We constantly read in the newspapers about cases of parental abuse. All too often it is a father who terrorizes his wife and children, sometimes over a period of many years. Often these family tragedies never come to light, but some cases gain publicity when one of the victims finally brings the matter to the attention of the authorities. The consistent failure of God to intervene in any way is disturbing. In many countries there are laws requiring any person who is aware of child abuse to report it; quite rightly, there are penalties if one fails to do so. The difficulty this raises in my mind is that, according to all of the mainstream religions, God must be aware of these problems; he seems, however, to do nothing. How can this inaction be justified? After all, if God insists on remaining in the background, he can easily intervene quietly, for example by giving the wife strength to take the necessary action. In some cases the wife has prayed to him over many years, and her prayers are simply not answered in time to avoid much psychological damage to the children. Often these children pass on the abusive practices to their own children. All of these problems would be avoided if God only did what the law requires us humans to do—to take the necessary action.

These difficulties have, of course, troubled people for a long time. They were referred to in the Old Testament book of Job, and were raised in the fourth century by St. Augustine. They were later brought out clearly by David Hume, in his *Dialogues Concerning Natural Religion* (1750). How, Hume asked, can we reconcile the idea of an omnipotent, all-knowing, and good god with the existence of evil in the world? If God were good he would take action to reduce the evil in the world. The fact that he does not do so must be either because he is unable to do so or is unwilling. In the former case he is not omnipotent; in the latter case he is malevolent. There seems to be no logical alternative.[9]

Those who face the problem of evil in this world must consider two forms, natural and moral. Natural evil, commonly (and rather ironically) referred to as acts of God, includes earthquakes, volcanoes, tidal waves, hurricanes, and diseases. It also includes the cruelties daily inflicted by wild animals on one another. Richard Dawkins, in *River out of Eden* and other books, has written eloquently about the sad fate of the antelopes, which are regularly torn to pieces by cheetahs. If one believes that these beautiful creatures were created by God, one is forced to conclude that cheetahs were meticulously designed to be highly successful at killing antelopes, while antelopes seem to have been designed to be highly effective at eluding cheetahs—in other words, at trying to starve them to death. Biologists seem to be more likely than scientists in other fields to be religious nonbelievers, and no doubt this is due to their greater awareness of these natural evils.

Moral evil includes all the acts of violence so widely inflicted by human beings—wars, murder, torture, and so forth. Some evil cannot be clearly identified as natural or moral. Hunger is one of these. Modern technology has made it possible to feed the present world population, but there is an inefficient distribution of food. The responsibility for this lies with many incompetent, indifferent, and evil human beings. Hunger also results, to a considerable extent, from overpopulation, partly because some powerful churches forbid (fortunately only with limited success) the use of the most efficient methods of birth control. Often human beings contribute to the suffering from natural evil by doing foolish things, like building cities on fault lines and then complaining when an earthquake occurs. I am told there is a motel that effectively advertises itself by proudly claiming to be right on a fault line!

It is particularly difficult to justify the existence of moral evil. Theologians respond that God has provided us with free will, which many humans misuse to choose evil. God does not intervene since that would be to deprive us of free will. This, however, is hardly a strong argument in favor of praying to God; indeed, it suggests that intercessory prayers should be dispensed with, since what is the point of praying to God if we know that he will not intervene? Also, the importance in modern prayer of expressing praise of God and gratitude to him seems to me incongruous in a world that is so full of misery. Religious bodies are inconsistent about these matters; how, for example, can it be justifiable to say, as they do in Christian churches, "The Lord is my shepherd, I shall not want"?

If praying to God had some effect, would there not be some evidence for it? In times of war, many prayers are offered from both sides, but no historian has ever produced evidence that the outcome of a war was influenced by the more effective praying of one side. Years ago Darwin's cousin, mathematician and statistician Sir Francis Galton (1822–1911), pointed out that insurance brokers offered no discounts to what he called the "praying classes." There is some evidence that people suffering from serious diseases such as cancer and AIDS respond a little better to treatment if they pray, but the small effects observed can be understood on the grounds that one's psychological condition affects one's physical condition. This is supported by the effectiveness of placebos. If there were any convincing evidence for the effectiveness of prayer, doctors would be prescribing prayers as well as medical treatment, and most patients would be praying as hard as they could.

There is statistical evidence that people who attend church regularly live longer and healthier lives than those who do not. We might draw the conclusion that God favors those who pray to him, but that would be naive. The result can be fully explained by the fact that people who go to church tend, on the whole, to live more disciplined and therefore healthier lives than other people; they are less likely to become addicted to drugs and alcohol, for example.

The theological answer to these difficulties, that God does not intervene in our free will, seems unsatisfactory for many reasons. Some people are fortunate enough to be able to handle situations well and to make something of their lives by the exercise of their free will. But it is unreasonable to assume that all humans are able to exercise their free will effectively. What is the use of free will to a small child suffering from parental abuse? What is the use of free will to the countless millions who are imprisoned for no valid reason? What is the use of free will to the many millions of hungry children who will starve to death during the next twelve months? What is the use of free will to a person born with a crippling genetic disability? Where is God while these people are suffering? If God were a good and just god, surely he would intervene in human affairs in a more effective way. It is because of questions like this that many thoughtful people today conclude that either God does not exist, or that he is not able or willing to take any action on our behalf; in either case, prayer can only be ineffective.

All available evidence leads to the conclusion that, subsequent to the creation of the universe at least twelve billion years ago, the laws of nature have been followed; there is no evidence for divine intervention at any later time. So praying to God and worshiping him seem pointless, and can be justified only in terms of a psychological advantage to oneself if one believes. If there is a God and an afterlife, it seems to me that things must be arranged in a way that is quite different from that maintained by conventional religions. Perhaps God admits into his heaven those of us who have led decent lives. If a conventional religious belief leads a person to a good life, so much the better for it. But I cannot believe in the kind of god who judges us by how much we flatter him in our prayers.

The arguments so far can be appreciated by anyone who looks at the world around us; they require no specialized scientific knowledge. Those with some knowledge of science—admittedly fragmentary, since there is so much that is still not understood—are aware of many more reasons not to believe in a personal god. The universe and living things have evolved in such an inefficient way, with chance playing such an important part, that it is hard to believe that an omniscient and all-wise god would have been satisfied with them. Many examples will be obvious from the discussions in earlier chapters. Why did God not create a suitable Earth for humans to live in, instead of making a universe dominated by so many inhospitable places? Why could he not have made a better Earth for us to live on, instead of one in which earthquakes, floods, and hurricanes cause so much misery for so many? Why was he not more selective in his creation of species—why cheetahs and antelopes, why 300,000 species of beetle? Why did the human race evolve from primitive species, ending as a highly inefficient being with many undesirable genes and so much luck in the way they are distributed? Why are so many things in our lives purely a matter of chance? We can hardly help concluding that if there was an act of creation, the job was hopelessly bungled, not what we would expect from an all-powerful god.

Over the years I have developed a personal philosophy of life in which supernatural events, divine intervention, and an afterlife play no part. To me, religious belief creates more intellectual problems than it solves.

There are many intelligent and thoughtful people who freely accept many of these arguments, but nevertheless believe in a personal god who answers their prayers. Such people often admit that the existence of so much evil in the world creates an intellectual difficulty. However, instead of rejecting the existence of a living god on the basis of these arguments, they find that their belief helps them to resolve their difficulties, and to lead more satisfying lives.

Such people, even if they accept the findings of modern science, have a sincere conviction that there is a god who created the universe and is actively concerned with our welfare. In other words, they have faith, while others of us have not. Our philosophy of life is mediated by our genes, and it also results from our environment—from following beliefs that were often introduced in early years. All religions contain several interrelating components. One of them is a historical account of events relating to the creed; in Christianity, the Old and New Testaments play this role. Another, and perhaps the most important, is a code of ethical behavior; in Christianity this is embodied in the Ten Commandments, modified somewhat in the New Testament and expanded and clarified by the teachings of individual Christian sects. Of course, as I have discussed, I do not include a cosmology as an essential part of a religion; religious bodies must leave cosmology to science.

Aristotle taught that human beings are differentiated from other animals by the gift of reason, and that we all have a duty to exercise to the full our moral and intellectual characters. Every reasonable person agrees that we should follow a moral code, but there is disagreement about its origin. Some believe that a moral code comes to us from God, and that one function of religion is to clarify and interpret God's moral code. Others—and this was Aristotle's point of view—take an empirical approach and think that the moral code comes from a thoughtful and sincere interpretation of the world around us, in particular of how the world can function to the good of all people. In many ways, the origin of moral codes does not really matter. Reasonable people, whether they believe in God or not, often arrive at essentially the same moral code. There are always disagreements about details, even between people belonging to the same religious denomination.

Some people hold strongly to their religious belief because it provides them with a moral code that they accept instinctively. In other words, they believe that the moral code they would formulate empirically corresponds to the one provided by their religion. Such people find themselves convinced that their moral code and their God are inseparable. Many nonbelievers, however, accept just as strict a moral code from personal choice, without any fear of retribution.

It is an empirical fact that religion has a strong attraction for many human beings. Believers derive comfort and satisfaction from their conviction that God is watching over them and listening to their prayers, and that there is a next world in which the injustices of this life will be rectified. Many people enjoy the comradeship of those who share their belief, and for them the discipline of attending religious services regularly is undoubtedly a salutary one. A religious creed provides answers to deep and troubling questions about human existence, answers that believers find satisfying.

Religious belief is often criticized by agnostics because it has led to much evil in the world. There is a little irony in this accusation, since, in recent times, the Communist movement, with its emphasis on atheism, has perpetrated comparable cruelties. The Nazi regime was also not based on religious belief. Admittedly, there are many instances where hatred has been inflamed by religious beliefs and even by religious organizations. The Crusades are often quoted as examples, as are the bitter persecutions of Roman Catholics by Protestants and of Protestants by Catholics. Even today there are many incidents of this kind, particularly in Northern Ireland, in Yugoslavia, and in many Muslim countries. Sincerely religious people deplore these actions as much as nonbelievers do. They argue, with some justification, that such actions do not arise from true religion but from the actions of evil people who use religion as an excuse for violence. Many of the Irish Protestants and Catholics who shoot one another in Ireland are not religious people at all—many of them never go to church—

but are lovers of violence who need an excuse for what they do. Decent Irish people are as critical of them as nonbelievers are of religion.

Every human activity has its good and its bad aspects, and, to be fair, we must consider both. Science is morally neutral, and as well as doing good it has led to some evils—the atomic bomb, the misuse of drugs, and so forth. We do not abandon science because of these unfortunate consequences, but instead take a balanced view. We recognize that all the advantages provided by science—rapid communication, rapid travel, an ample supply of pharmaceutical drugs, and anesthetics, to name a few—should not be thrown away because of the sometimes evil consequences that can come from their misuse.

So it is with religious belief. We must consider its good and bad consequences and strive to make the former prevail. Religions, or rather aberrations of religion, have started wars, but sometimes religious leaders have mediated between quarrelling nations and prevented wars. Wars get the publicity; newspapers and history books do not report wars forestalled by religious bodies. Religious organizations have played an effective and honorable role in other events. The collapse of the Soviet Communist system in Eastern Europe owed much to the influence of the Roman Catholic Church, especially to its leader, Pope John Paul II. Apartheid in South Africa was defeated with little bloodshed partly because of the efforts of Anglican Archbishop Desmond Tutu. In other parts of Africa, Archbishop Trevor Huddleston played an equally distinguished role. I think it fair to say that organized religions at the present time do more good than evil; most of the evil is done by aberrant members of religious bodies, such as Christian or Muslim fundamentalists.

Religious organizations have also done good by feeding and tending the poor and running hospitals. In former times, governments did little social work, today regarded as essential. The churches assumed the entire responsibility. Many hospitals that are now run by governments were begun by religious organizations, as some of their names remind us: for example, London's St. Bartholomew's Hospital, where Sherlock Holmes is supposed to have met Dr. Watson. The same is true of charitable organizations such as orphanages. It would be no exaggeration to say that many of our modern hospitals and other social services belong to a strictly religious heritage.

Organized religion also has an honorable record for education and scholarship. Many of the universities in the Western world are Christian foundations, while the earlier Arab universities were Islamic foundations. Some older universities, like Oxford and Cambridge, were Roman Catholic foundations, becoming Protestant at the Reformation, and with religious control remaining until the nineteenth century. It is true that the Anglican control of these universities (notably at Oxford and Cambridge) was somewhat abused by the exclusion of teachers and students who did not conform. In spite of that, the tradition laid down during those earlier years was influential and long-lasting. Most modern universities, although they never had any religious control, are organized in a similar manner to the older foundations, so that the educational methods established by church authorities are still recognized as effective.

Religious bodies have played a beneficent role in the worlds of scholarship and science. Admittedly, there were unfortunate lapses, such as the treatment of Galileo and the Roman Catholic Church's suppression of the teaching of science in the nineteenth century. However, these incidents tell us more about the deficiencies of religious leaders in times of stress than about religion itself. There are many more examples where religions have played an enlightened role in scholarship and science. The establishment of great libraries and the development of the fields of philosophy and history were for many centuries the work of religious bodies. The same is true of scientific research.[10] The foundation of the Royal Society in seventeenth-century England was brought about by a group of conscientious Anglicans, many of whom were clergymen. Anglican clergy also played a prominent role in the 1831 founding of the highly effective British Association for the Advancement of Science; of its first fourteen presidents, six were clergymen.

Many clergymen and devout laymen have made substantial contributions to the development of science. Among laymen who took their religious beliefs seriously were Robert Boyle, John Dalton, Clerk Maxwell, Michael Faraday, Lord Kelvin, and Sir Arthur Eddington. Among ordained clergy of various denominations was the Reverend Roger Joseph Boscovich, a Jesuit priest whose atomic theory and other scientific contributions in the eighteenth century exerted a wide influence, and who remained in good standing with his church. Another example is Unitarian minister Joseph Priestley, who did pioneering work on the chemical properties of gases. The Reverend Gregor Mendel did notable work on genetic selection, and the Abbé Georges Lemaître did important work in cosmology.

At the same time, it has to be admitted that the attitude of some religious bodies to science has been less than enlightened. This was particularly true in the past of the Roman Catholic Church. For example, the works of Copernicus were placed on the *Index Librorum Prohibitorum* (List of Prohibited Books), and Galileo was ordered not to follow the Copernican theory that the Sun is at the center of the solar system. In 1635 Galileo was summoned to abjure his "heresies," and his ideas about the solar system were officially condemned by the Roman Catholic Church until 1922.

After the publication of Darwin's *Origin of Species* in 1859, the position of the Vatican became even more repressive. In 1864 Pope Pius IX issued a blanket condemnation of science in general, and the teaching of science was strictly limited in Catholic schools and universities. Catholics who favored the theory of evolution were excommunicated, and their writings placed on the *Index.*

Fortunately, there has recently been a relaxation of these condemnations. In 1982 the pontifical Academy of Sciences, which advises the Vatican on scientific matters, issued a statement that "masses of evidence render the application of the concept of evolution to man and the other primates beyond serious dispute." Finally, in 1996, Pope John Paul II issued a statement accepting this point of view.

One fact that has to be taken into account in any discussion of religion is that there has undoubtedly been a marked decline in religious belief during the twentieth century. Many denominations have reported that attendance at religious services has steadily decreased. Unhappily and astonishingly, however, some fundamentalist churches in the United States report an increase in attendance.

It is important, of course, to distinguish between religious belief and church attendance. Many people attend church regularly but have no religious belief at all. They attend for a variety of reasons: it is socially beneficial, it makes a good impression on their employers, or they enjoy the comradeship. Also, there is a natural human tendency for people to cling to their previous beliefs; people are more comfortable with ideas instilled at an early age.

Many people adhere to a religious body and give it full support, while at the same time having serious reservations about their belief in God. An outstanding example is William Temple (1881–1944), who was Archbishop of Canterbury from 1942 to 1944. He said that when he recited the Christian creed ("I believe in God, etc.") he mentally prefaced it with the words "I am prepared to lead my life as if. . . ."[11] Essentially, he was redefining religious belief as the willingness to behave *as if* one believed in the teachings of the church. Some would regard this attitude as hypocritical, but William Temple (the son of Frederick Temple, who was mentioned on page 186) was a learned and good man who presumably had reflected sincerely before adopting his position. Many others who see problems with their religious belief remain regular churchgoers, but take an "a la carte" attitude, picking and choosing those parts of their religion that they find convenient.

Many people who do not believe in God and never attend a place of worship will, if they are asked, insist that they believe in God, because they think it is the "respectable" thing to do. So I think that surveys of religious belief tend to overestimate the number of true believers. More reliable information comes from records of church attendance, which has clearly declined over the last few decades.

Religious belief in people who have distinguished themselves by intellectual pursuits is especially interesting. Surveys have been made over a period of many years of members of the U.S. Academy of Sciences, one of the most distinguished and exclusive scientific academies in the world. A survey of academy members taken in 1914 showed that 27.7 percent of the respondents said that they believed in a personal god. By 1933 the percentage had gone down to 15, and in 1998 the percentage was only 7. There was a corresponding disbelief in the possibility of human immortality. Similar declines have been revealed in other surveys of members of professional groups.

In contrast to this, surveys of members of the general public reveal a somewhat larger, although declining, acceptance of religious beliefs. It is difficult to escape the conclusion that there is something of an inverse correlation between intelligence and education on the one hand, and religious belief on the other. An uneducated person is more likely to be religious than a university professor. Of course, the correlation is statistical only; some highly intelligent and well-educated people are religious, and some uneducated people are nonbelievers.

Just what conclusions can be drawn? One is that religious belief cannot be entirely a matter of heredity, since genetic factors do not bring about changes in human nature in such a short time; several centuries would be required for the kind of change that the U.S. Academy survey revealed from 1914 to 1988. Changes in our culture must also play a role.

Whether or not we are religious in a conventional sense depends on our particular genes and memes. I explain my own position as an nonbeliever by concluding that I was born with a set of genes that did not give me a strong predisposition toward any yearning for a spiritual life. As a boy I was taken regularly to an Anglican church and I attended a school and university where the Anglican tradition was strong. At no time, however, was I coerced to believe in God, and I remember being a skeptic at an early age. When I was first introduced to the biblical story of Doubting Thomas, who demanded "a sign," my conclusion was that he was the only intelligent disciple; the others were gullible and he very properly required some evidence.

Apparently I was born with the cast of mind that leads to intellectual pursuits, particularly science, rather than to religion. Intellectual curiosity and the love of intellectual pursuits are strongly linked to skepticism and the refusal to accept things without question. One of the most important aspects of higher education is that one is encouraged to question traditional beliefs and opinions. Research, in science and indeed in any field, is done most successfully by those who question previously held opinions. Also—and I think this is particularly important—the understanding of the universe and life that is provided by the latest scientific knowledge seems to me more intellectually satisfying than that provided by any religion. In this I am in agreement with Paul Davies, who, in several penetrating books, has discussed the latest scientific ideas in relation to religious belief. In *God and the New Physics* he stated that science offers a surer path to God than religion. By God, in this context, he meant an understanding of our place in the universe.

After a life of scientific research I have become convinced that the only reliable way to approach the truth is to gain information, by observation and experiment, about the universe around us, and then apply rational and coherent arguments to that information. Any other procedure, like relying on authorities or using intuition, seems treacherous. Imagination and intuition are certainly helpful in leading to ideas, but in the long run we must always rely on empirical tests. This is why I find science much more reliable than religion.

The more I have thought about religion, the more dissatisfied I have become with all of the recognized religions. The main problem is that they seem to have created God with many of the characteristics of a human being. But if there *is* a God, surely he is entirely different from us. I can accept the idea of a god who is a superb mathematician and physicist and has established the laws of nature for us. But having done that, he leaves us strictly alone, to our own devices. All the evidence from the world around us, such as from quantum and chaos theories, is that he leaves everything to chance, never intervening in any way with the course of nature.

Religious people have sometimes commented to me that I should be unhappy after so many years without feeling any support from God and without any hope of life after death. My point of view is just the opposite. It seems to me that believing in God raises many more intellectual difficulties than it solves. I find it impossible to rationalize a world that is in any way ruled by God; far too many questions remain to which I can see no reasonable answers. That the world is controlled entirely by the laws of nature, including the laws of chance, seems to explain everything much more satisfactorily. Living by ethical guidelines based on a sense of right rather than fear of retribution causes me no philosophical difficulties.

*   *   *

We are all very much the slaves of our memes and genes. We are what we are for reasons that are largely beyond our control. I admit that I may be all wrong; if so, I may discover the truth in an afterlife, and have to use a retort like Bertrand Russell's, quoted at the beginning of this chapter. I would express it a little more strongly because if God exists, he did more than just make the evidence for his existence insufficient. He has included such strong evidence for his failure to take any action on behalf of human beings that he has inevitably amplified disbelief among those of us who have spent so much effort in trying to understand the secrets of his universe.

Many of us find troublesome the lack of coherence in organized religious belief. In *Science and Wonders* (1996), Russell Stannard reports discussions with a variety of people about such matters as the origin of the universe, the origin of life, and the workings of the mind.[12] Those who professed to believe in God (including a few scientists) gave a bewildering variety of answers, for the most part quite incompatible with each other. The nonbelievers, on the other hand, presented a largely coherent picture, with disagreements only on trivial details. The divergences between the believers were all the more striking since they were nearly all Christians; if representatives of other religions had been consulted the overall result might have been even more incoherent.

I cannot help concluding, however, that the world would be a better place if we all based our opinions not on the teachings laid down centuries ago by religious bodies but on an unprejudiced assessment of what scientific investigations have revealed. Many religious institutions regard themselves as the sole exponents of absolute truth, and have become sharply critical of one another, sometimes with deadly results. Religion is too easily perverted by fanatics.

Science and other scholarly pursuits, on the other hand, make no claim to having attained absolute truth, merely to approaching the truth. It is true that science has from time to time been perverted, and that its products, like atomic bombs and other weapons, have been used for undesirable purposes. It is hard, however, to believe that terrorism, so often carried out in the name of religion, would ever be carried out in the name of science. Since terrorism seems to have become the main scourge of society, we now have an additional reason for displacing religions, especially authoritative ones, in favor of a rational philosophy of life based on the observational and experimental evidence derived from the universe around us.

Chapter 8
What Is Truth?

What is truth? said jesting Pilate, and would not stay for an answer.

—Francis Bacon, "On Truth," 1625

It may be true that on that memorable occasion Pontius Pilate would not stay for an answer, but I doubt that he was jesting. More likely he was awed by the difficulty of his task, as we all must be in seeking the truth. By its very nature, the truth is something we can look for, but we can never be sure that we have found it. That, at any rate, is the position that I believe most scientists take, and I think it applies to all different kinds of truth.

We must recognize that there are several different kinds of truth.[1] Sometimes artistic people use the word truth when they talk about a painting, a poem, or a work of fiction; that is aesthetic truth. Other kinds of truth are religious truth, legal truth, historical truth, and scientific truth. It is unfortunate that the same word, "truth," is used since each meaning is different. Religious truth is very different from other kinds, since faith plays a large role and evidence a negligible one.

In many of the other kinds of truth, the perceived facts are of dominant importance. In a court of law, a witness is required to speak "the truth, the whole truth, and nothing but the truth." Witnesses are usually required to stick to the "facts," meaning the facts perceived by them. Witnesses, even if they are intelligent and honest, often give mutually inconsistent accounts of their perceptions of the facts. At the end of a trial, a judge summarizes the evidence and interprets it, and then the jury reaches a verdict. It is hoped that the verdict corresponds to the truth, and no doubt it sometimes does. Occasionally it emerges later that the verdict was far from the truth. Under ideal conditions, in a well-conducted trial, we may be reasonably confident that the verdict is close to the truth, but we cannot be certain. Certain legal criteria have been applied, hopefully without serious error. As in all kinds of truth, we just have to do the best we can.

With historical truth the situation is somewhat similar, except that we are concerned more with documentary evidence and less (except for recent history) with evidence from eyewitnesses. Again, we sift through the evidence and come to what we think is a reasonable conclusion.

There are two essentially different ways of approaching the truth. The method used in the early years of civilization was the *intuitive* method—the truth we feel instinctively. Early humans believed that the earth was flat and that the Sun went around us. In the Old Testament, for example, this view is taken for granted, and it took many centuries for it to be corrected. The intuitive method was largely used by Aristotle, who still ranks as a scientist of great distinction, even though he sometimes went badly astray. He felt that certain things *must* be true, and often did not bother to test them. As a result of his dependence on intuition, Aristotle was less successful in physics, where theory is always important, than in biology, which was then a descriptive science.

The intuitive method continued to be the most important method until the Renaissance, when more emphasis began to be placed on the *empirical* method. This is the method that scientists depend on for their final conclusions, as do many other people in a variety of situations. This method emphasizes observation, experiment, and other kinds of evidence. The essential feature of the method is that it *works in practice*, and thus is the best method we have so far.

Intuition is still important, as long as it is used wisely. The *Shorter Oxford Dictionary* gives as its first definition of intuition, "immediate apprehension by the mind without reasoning," but I think we need further clarification and consideration. Agatha Christie's detective, Hercule Poirot, sometimes makes interesting comments about intuition. In *The A. B. C. Murders*, for example, he remarks, "In a well-balanced mind there is no such thing as an intuition—an inspired guess," and later he says "But what is often called an intuition is really based on logical deduction or experience. When an expert feels that there is something wrong about a picture or a piece of furniture . . . he is really basing that feeling on a host of small ideas and details." In other words, intuitive ideas spring to one's mind on the basis of background knowledge and previous thinking. Poirot went too far in implying that a rational person never has an intuitive thought. The point is rather that such a person treats intuitive ideas in a critical but constructive way, bringing the light of reason to bear on them. That is, there is nothing wrong with using intuition for preliminary thinking, but one must always use the empirical method to reach a final conclusion.

It is thus an oversimplification to say that in science and other scholarly fields empirical methods have *entirely* taken the place of intuitive ones. On the contrary, scientists make good use of intuition, and some highly creative scientists are mainly intuitive in their approach to research. As a graduate student at Princeton, I worked with Henry Eyring, a chemist of distinction who is remembered for his formulation of a useful and original theory of the rates of chemical reactions. Eyring was always bubbling with ideas, most of which seemed to come from thin air. On closer examination, most of them turned out to be wrong, as he would usually readily admit; occasionally he was hard to persuade. Out of all the ideas he had, a few turned out to be right; when developed and compared with experiment (the empirical method), some turned out to be very important.

Nonscientists often criticize scientific theories on the grounds that they seem unreasonable, meaning counterintuitive. We have seen examples of this in earlier chapters. Einstein was criticized by philosophers on the grounds that his theory of relativity defied their intuitive ideas about space and time. Planck's quantum theory was criticized, even by some eminent scientists, on the grounds that the idea of packets of energy seemed unreasonable. The basic error these critics made was to use intuition as a valid way of testing a scientific theory. It is essential to realize that intuition may be of great help in formulating a theory, but is thoroughly unreliable for testing it.

Reliance on authority is sometimes regarded as a path to the truth, but it is a derived path; we rely on someone else to formulate our ideas. Religions often require one to accept without question the authority of their leaders. In practice, scientists also depend to a great extent on authority, perhaps more than some of us would like to admit. The quantity of scientific information is now so vast, and the theories so intricate, that no scientist today can hope to verify much. No one could possibly check everything that appears in the scientific literature, even within a narrow branch of science. We have to depend to a great extent on the authority of others, assuming tentatively that what we read is correct—in other words, we have to have a certain amount of faith in others. Of course, we are always alert to the fact that we may be reading something fraudulent or nonsensical. This is perhaps what distinguishes science from religion more than anything else; with religion, one accepts a belief without question, while in science one is always skeptical.

Einstein's famous formula, $E = mc^2$, provides an interesting example of the temporary reliance that scientists sometimes place on authority. The equation appeared in 1905, but convincing experimental evidence for it only emerged in the early 1930s. During the intervening period scientists made some use of it, but kept in mind that it might be proven wrong. Some of the latest theoretical ideas, such as superstring theory, are in a similar situation; there is still no convincing evidence for them, and scientists treat them with caution.

I have emphasized that there is no such thing as a scientific method, that is, a method used only by scientists. Instead, there is a method used by scholars in all fields, and also by medical practitioners, lawyers, engineers, policemen, and plumbers. This method I have called the judicial or academic method. Several important criteria are essential to this method. One is *objectivity*: we should consider only the relevant evidence and put aside our personal feelings and prejudices. A juror in a trial is supposed to arrive at a judgment entirely on the basis of the evidence presented at the trial. A juror should not be influenced by the color of the accused person or any other irrelevant factors. The same is true of a scientific investigation, which should be concerned only with valid information.

Another important criterion is *coherence*. Conclusions should be internally consistent; everything should hang together. This is particularly true of a complex legal judgment, which must be logically self-consistent. Otherwise the judgment simply would not *work*; it would be subject to widespread criticism and lead to confusion. Exactly the same is true of a scientific publication and a medical report.

Third, any approach to the truth must be recognized as *provisional* and, to some degree, *approximate*. A scientific theory, a medical report, or a legal judgment may be the best that can be done in the circumstances, but it may well be wrong. It may be badly wrong, as in the case of people who are wrongly convicted or whose ailments are incorrectly diagnosed. We can approach the truth, but we can never be sure of reaching it.

We should consider these three criteria a little further in relation to a scientific investigation.[2] Scientific ideas have to be objective because they are open to the scrutiny of anyone interested. The convention in science is that a piece of research is not officially recognized until it has been published in a scientific journal. Normally a paper is not accepted for publication until it has been given some scrutiny by the editor of the journal or his delegate, who acts as a referee; this has come to be known as peer review. To get a scientific paper published, we are required to include enough detail that readers are in a position to repeat the experiments and check the mathematics.

This convention is an important safeguard, since errors in the research are soon discovered. The alleged discovery of "cold fusion," discussed in chapter 2, is a good example. The work was first announced to the press, a violation of scientific etiquette; it should have been submitted for publication in the conventional way. After the details of the research were revealed, its unreliability was discovered as soon as experiments were adequately repeated. Because of certain complications, this took a few years, more than usually necessary.

The objectivity of science has come under attack, particularly from the "postmodern positivists" and "cultural relativists" (chapter 6). Pointing to some examples in which scientists have not been objective, they try to discredit all of us. Scientists are well aware of these mistakes and have invented safeguards. The claim of some critics that science is no more than a social construct, which depends largely on the current fashions of thought, is manifestly absurd. Science *works*, and leads to things that work. Devices like televisions and computers would hardly work if the science had no validity. A particularly bizarre form of cultural relativism is found in the opinions of a minority of feminists who argue that, since men have in the past played a more important part than women in the advancement of science, science somehow has a masculine stamp on it, and would be very different if women had been more involved. Would the laws of nature really have been different if women had discovered them? I have never heard this point of view expressed by any of the distinguished women scientists I have known, or by any of the excellent women students I have had. Nature seems to work just the same for them. However, the industrial application of science might well have followed a more humane path.

Coherence in science means that it must hold together and form a self-consistent and logical network. We saw this clearly in chapter 5, where we considered the origin of the universe and life on Earth. Results from experts in many diverse sciences—physics, chemistry, astronomy, geology, and biology—lead to a consistent conclusion about the time scale in which these events occurred. Scientists who are at the forefront of knowledge have to be reasonably well informed about relevant aspects of sciences other than their own. An astronomer or biologist who knows little physics or chemistry does not make many important discoveries. Perhaps most important, all scientists must be reasonably proficient in mathematics.

Science is a complex network of laws of nature, models, and theories. As a result, the whole structure is much stronger than any single component of it. There may be some weak parts of the structure, but the rest of it is strong enough to hold it together. Eventually the weak components will be identified and replaced or repaired. A good example of repair is provided by the theory of relativity. It did not completely replace Newton's laws of motion, which are still serviceable for most purposes. Instead, it adjusted and elaborated them to apply to the special case of objects moving at enormous speeds, and dealt with a few other problems outside the scope of Newton's equations.

Scientific knowledge is approximate and provisional. Anyone who has studied the history of science is very conscious of this. At various times scientists decided that most of the problems of science had been solved. Even as great a scientist as Lord Kelvin fell into this error toward the end of the nineteenth century, suggesting that only a few trivial details of physical theory needed to be worked out, and then everything would be understood. In the future, he thought, there would be not much for scientists to do besides make more accurate measurements. Kelvin never suspected that great developments would spring from the discovery of radioactivity and x-rays, and that two comprehensive theories, of quanta and of relativity, would lead to fundamental changes in our scientific thinking, and so to remarkable scientific and technical advances.[3]

The scientific events of the twentieth century have made us wary of such predictions. In *What Remains to be Discovered*, Sir John Maddox makes predictions about science that are in striking contrast to Kelvin's, and are overwhelmingly convincing. Far from concluding that almost everything has been solved, he emphasizes that the number of unanswered questions increases rather than decreases as discoveries are made. To explain this apparent paradox, Maddox introduces the intriguing analogy of unpacking a nest of Russian dolls. When we unscrew the outermost one, we discover another; when that is unscrewed, another is revealed, and so on. Unscrewing the early theory of the atom revealed electrons; further unscrewing discovered radioactivity, with its promise that atoms are not indivisible. Further unscrewing revealed neutrons and other subatomic particles. In a broadcast on 1 October 1939, Winston Churchill made a comment about the behavior of Russia that could also apply to scientific knowledge: "It is a riddle wrapped in a mystery inside an enigma."[4]

*   *   *

In this book we have looked at methods we can use to gain an understanding of our universe, and which are reliable for drawing conclusions about the many issues that confront us in our everyday lives. For scientific work, by far the best method is the one I have called the judicial method, in which we rely entirely on logic, objectivity, and coherence. This method has been remarkably successful in leading to a self-consistent and practical understanding of the workings of nature. It has brought together, from a variety of independent sources, compelling evidence for the ages of the universe and many of the structures within it, such as the earth and living things on the earth. The validity of the scientific method is further supported by the technological advances made possible by the advance of pure science.

Is there any reason why we should not use the judicial method in dealing with problems other than scientific ones? My own conclusion is that we should always use it in reaching our final conclusions. The judicial method is already used widely by scholars in fields other than science. They too carefully consider the evidence available to them in an unprejudiced manner, and draw logical conclusions. The same is true of those concerned with practical matters.

By contrast, the intuitive method, which comes to us instinctively and was used exclusively in earlier times, is often unreliable. It is helpful in our preliminary thinking, and many scientists have made good use of it. When faced with arriving at a final conclusion, however, they find it essential to test their ideas by the judicial method, paying due regard to the observational and experimental evidence.

Although intuitive methods are no longer relied upon by most scholars and practical people, they are still widely used in other areas. In particular, politicians and senior government bureaucrats are still largely wedded to intuitive thinking, and, to make matters worse, they also blindly adhere to the adversarial system. Surely the world would be a much better place if politicians would put their prejudices aside—particularly their allegiance to particular political parties—and try to judge issues in an unbiased way.

The great scientific advances of the last two centuries—quantum theory, relativity theory, and the mechanism of evolution—lead us to a new appreciation of the universe and our place in it. They offer a deeper understanding of all creation and better ways of dealing with practical matters. As the veil of ignorance is gradually being lifted by the advance of science, many including myself are being led to a philosophy of life that we find more rewarding than any other.

Notes

Chapter 1: To Tell the Truth

1. Pierre Teilhard de Chardin, *The Phenomenon of Man*, trans. Bernard Wall (New York: Harper & Row, 1975); originally published as *Le phenomene pumain*.

2. Thomas Babington Macaulay, "History," in *Essays* (New York: Sheldon, 1860), p. 387.

3. The physicists to whose tennis Nernst objected were F. A. Lindemann (1886–1967), later Lord Cherwell, and his brother C. A. Lindemann, later Brigadier Lindemann.

Chapter 2: The Nuts and Bolts

1. For more detailed accounts of the Blondlot affair see chapter 6 of Gratzer, *The Undergrowth of Science: Delusion, Self-Deception, and Human Frailty* (Oxford: Oxford University Press, 2001), pp. 1–28; and his *Eurekas and Euphorias: The Oxford Book of Scientific Anecdotes* (Oxford: Oxford University Press, 2002), pp. 75–78.

2. For further details of the cold fusion affair see chapter 6 of Walter Gratzer, *The Undergrowth of Science*, pp. 111–35. For a detailed scientific evaluation of the experiments see John R. Huigenza, *Cold Fusion: The Scientific Fiasco of the Century* (Oxford: Oxford University Press, 1993).

3. The signals were extremely faint, consisting of clicks against a strong background of static. One of Marconi's assistants was unable to detect them, so there remains some doubt that they were really received. Marconi's claim in 1900, however, undoubtedly led to the conclusion that radio waves could be sent across the Atlantic by being reflected off the ionosphere.

The publicity around Marconi led the public to think he invented radio. As I explain in my *To Light Such a Candle* (Oxford: Oxford University Press, 1998), Marconi made no technical innovations to radio transmission, always using equipment developed by others. The most important scientific contributions to radio broadcasting were, in my opinion, made by Heinrich Rudolph Hertz (1857–1894), who first transmitted radio waves in 1887, and by Sir Oliver Lodge (1851–1940), who developed improved equipment to generate and detect radio signals, and made important public demonstrations of the technique in 1889 and in 1894. It was only in 1896 that Marconi made his first experiments, using techniques similar to Lodge's. Marconi's contributions to radio were entirely on the commercial side.

4. *Report of the Trial of Prof. John Webster Indicted for the Murder of Dr. George Parkman, before the Supreme Court of Massachusetts, holden at Boston on Tuesday March 19, 1850*, phonographic report by Dr. James W. Stone (Boston: Phillips, Samson, & Co., 1850). The word "phonographic" refers to a system of shorthand or stenography, introduced in about 1837 by the English educator Isaac Pitman (1813–1897).

5. For further details about philosophers' early criticisms of relativity theories, see Gratzer, *Eurekas and Euphorias*, pp. 94–97.

Chapter 3: The Ingredients of Our Universe

1. "The filament would stretch to the Moon and back over twelve million times": The calculation is as follows:

One liter of water $= 55.5$ moles $= 55.5 \times 6.022 \times 10^{23}$ molecules
$$= 3.3 \times 10^{25} \text{ molecules}$$

In a chain of water molecules, held together by hydrogen bonds, the distance between neighboring oxygen atoms is about
$$300 \text{ pm} = 3 \times 10^{-10} \text{ m}$$

This is the length of each link in the chain, the total length of which is therefore
$$3.3 \times 10^{25} \times 3 \times 10^{-10} \text{ m} = 10^{16} \text{ m}$$
$$= 10^{13} \text{ km}$$

(This is a little more than a light-year, which is 9.462×10^{12} km.)

The distance to the Moon is about 4×10^5 km (it is about 1.3 light-seconds from us), so that the distance there and back is 8×10^5 km. The chain of H_2O molecules would thus go to the Moon and back
$$10^{13}/8 \times 10^5 = 1.25 \times 10^7 \text{ times}$$

or 12.5 million times.

When I first worked this out, the magnitude of the answer surprised me, so I made the calculation in a different way. Imagine the liter of water contained in a cube with sides of one decimeter (a decimeter is a tenth of a meter; one liter = 1 $dm^3 = 10^{-3}$ m^3). In the chain of water molecules the cross-sectional area is known to be about 10 square Angstroms; an Angstrom is 10^{-10} m, so the area is about 10^{-19} m^2. The length of the chain is thus

$$\frac{10^{-3} \ \text{m}^3}{10^{-19} \ \text{m}^2} = 10^{16} \ \text{m} = 10^{13} \ \text{km}$$

which is what we got by the other method of calculation.

2. The experiments by Benjamin Franklin of spreading oil on water are described in Charles Tanford's *Ben Franklin Stilled the Waves* (Durham, NC: Duke University Press, 1989).

3. For a much more detailed discussion of energy, see my *Energy and the Unexpected* (Oxford: Oxford University Press, 2003).

4. Chaos is discussed in more detail but in simple language in chapter 10 of my *Energy and the Unexpected.*

5. Ivar Ekeland, *Mathematics and the Unexpected* (Chicago: University of Chicago Press, 1988), p. 68. The effect of gravity on a billiard cannon was analyzed by M. Berry, "Regular and irregular motion," in *Topics in Non-Linear Dynamics,* American Institute of Physics Conference Proceedings, no. 46 (Washington, DC: American Institute of Physics, 1978), pp. 111–12.

Chapter 4: Our Place in the Universe

1. Henry David Thoreau, diary, 11 November 1850.

2. The story of the Piltdown man is well summarized, with references, in Walter Gratzer, *Eurekas and Euphorias: The Oxford Book of Scientific Anecdotes* (Oxford: Oxford University Press, 2002), pp. 204–207.

3. How seriously Martians were taken at one time is shown by the conditions for the Guzman Prize, established in 1900 by a French foundation. The prize of 100,000 francs was offered for the first contact with an extraterrestrial living being other than a Martian, which apparently would have made things too easy.

4. Newton described these experiments in his *Principia Mathematica;* see I. Bernard Cohen and Anne Whitman, trans., *The Principia: A New Translation* (Berkeley University of California Press, 1999), pp. 756–61. They were conducted in 1710 by the Reverend Dr. John Theophilus Desaguliers (1683-1744), who was for a time Curator of Experiments for the Royal Society of London.

5. If the universe is twelve billion years old, the galaxy has therefore traveled ten billion light-years in two billion years. Has it gone five times as fast as the speed of light? The answer is provided by Einstein's special theory of relativity: The galaxy cannot travel faster than the speed of light, but must have traveled at nearly the speed of light. Under those conditions the time frame is altered, in the sense that a fast-running clock runs slowly and can travel more than one light-year for every year that it records. According to the equation for Einstein's special theory of relativity, a clock on a galaxy traveling at 98 percent of the speed of light will travel five light-years but only record that it has traveled for one year. It therefore appears to have been traveling five times as fast as the speed of light.

6. James Hutton, *Theory of the Earth, with Proofs and Illustrations* (n.p., 1785).

7. Richard Dawkins, *The Blind Watchmaker* (London: Longmans, 1986; reprint, London: Penguin, 1988; reprinted with an appendix, 1991), p. 116.

8. "The neat packing required to get twenty-three pairs of chromosomes into the nucleus": The nucleus of a human cell is roughly one micrometer in diameter; to make it one meter across requires a magnification of about one million. The length of a DNA chain in one chromosome is about five centimeters (two inches), and its thickness roughly one nanometer (10^{-9} m, or about four billionths of an inch). Magnification by one million gives:

Length = 5×10^{-2} m $\times 10^{6}$ = 5×10^{4} m = 50 km (about 30 mi.)

Thickness = 10^{-9} m $\times 10^{6}$ = 10^{-3} m = 1 millimeter (mm) (three hundredths of an inch)

Chapter 5: How It All Began

1. The discrepancies between the findings of science and some religious teachings are brought out particularly clearly in Martin Gorst, *Measuring Eternity: The Search for the Beginning of Time* (New York: Broadway Books, 2001).

2. Sir Martin Rees, *Before the Beginning: Our Universe and Others* (London: Touchstone, 1997).

3. Here are a few more details about the value of the Hubble constant. When measurements were made from observatories on Earth, even at different positions in its orbit, the values obtained for the constant were rather widely spread, although all pointed to an age of the universe in the billions of years. Recent measurements made by astronomical satellites, particularly the European Space Agency satellite *Hipparcos*, provide more reliable values. Determinations made in 1999 and later range from about fifty to seventy-five kilometers per second per megaparsec (which can be written as km s-1 Mparsec^{-1}).

This unit, usually used by astronomers for the Hubble constant, requires the following data:

1 parsec = 3.262 light-years; 1 Mparsec = 3.262×10^{6} light-years

Speed of light = 2.998×10^{8} m s^{-1}; 1 year (y) = 3.156×10^{7} s

1 light-year = 2.998×10^{8} m s^{-1} $\times 3.156 \times 10^{7}$ s

$= 9.462 \times 10^{15}$ m $= 9.462 \times 10^{12}$ km

1 Mparsec = $3.262 \times 10^{6} \times 9.462 \times 10^{12}$ km

$= 3.087 \times 10^{19}$ km

A Hubble constant of 1 km s^{-1} Mparsec^{-1} is thus:

$(1/3.087 \times 10^{19})$ s^{-1} = 3.239×10^{-20} s^{-1} = $3.239 \times 10^{-20} \times 3.156 \times 10^{7}$ y^{-1}

$= 1.022 \times 10^{-12}$ y^{-1}

The two extreme values quoted above therefore correspond to

50 km s^{-1} Mparsec^{-1} = $50 \times 1.022 \times 10^{-12}$ y^{-1} = 5.11×10^{-11} y^{-1}

75 km s^{-1} Mparsec^{-1} = $75 \times 1.022 \times 10^{-12}$ y^{-1} = 7.67×10^{-11} y^{-1}

If we take the second value, and assume the Hubble constant to have remained the same since the big bang, the time that has elapsed since the galaxies started to move apart is

$1/7.67 \times 10^{-11}$ years = 1.30×10^{10} years = 13.0 billion years

The first, lower, value of the Hubble constant leads to 19.6 billion years. The following table relates Hubble constants to ages:

H/ km s^{-1} Mparsec^{-1}	Age of universe / billion years
50	19.6
60	16.3
70	14.0
75	13.0
80	12.2
100	9.8

The most recent work favors the lower values of the Hubble constant, which would suggest an age of eighteen to twenty billion years. However, because of the gravitational attractions between the galaxies, the constant must have been bigger in the past than it is today. The age of the universe is therefore smaller than obtained from the modern measurements of the constant. The best theoretical treatments of the correction required from this gravitational effect suggest that the calculated ages should be reduced by 30 percent. Thus, instead of an age of 19.6, we assume a more realistic value to be 13.7 billion years.

The value of twelve billion years that I have used in the main text is therefore a conservative one. For further details see Martin Gorst, *Measuring Eternity: The Search for the Beginning of Time* (New York: Broadway Books, 2001).

4. See, for example, the books by Atkins, Gribbin (*In Search of the Big Bang*), Silk, and Weinberg, listed under Suggested Reading.

5. "From the characteristics of the radiation": The way the intensity of radiation varies with the frequency is different for each temperature; therefore, a study of the frequency distribution allows us to determine the temperature of the source that is emitting the radiation.

6. The amount of deuterium, D or $_1^2$H, in the universe is actually about one ten-thousandth of the amount of $_1^1$H. This is very little compared to $_1^1$H, but there are more D atoms in the universe than atoms of either carbon or iron.

7. In fact, Darwin rather overestimated the age, now considered to be more like 135 million years. This caused him some embarrassment, as discussed in Gorst, *Measuring Eternity*, pp. 165–70, 303.

8. See John Campbell, *Rutherford: Scientist Supreme* (Christchurch, New Zealand: AAS Publications, 1999), pp. 279–81.

9. Paul Davies, *The Fifth Miracle: The Search for the Origin and Meaning of Life* (New York: Simon & Schuster, 1999).

10. The conditions for the evolution of higher forms of life are so stringent that it is possible that we humans are unique in the universe. This is argued very persuasively by Peter D. Ward and Donald Brownlee in *Rare Earth: Why Complex Life Is Uncommon in the Universe* (New York: Copernicus, 2000).

11. Richard Dawkins, *River Out Of Eden* (London: Weidenfeld & Nicholson, 1995).

Chapter 6: Science and Culture

1. This view is persuasively argued, for example, by Kenneth S. Deffayes, *Hubbert's Peak: The Impending Oil Shortage* (Princeton: Princeton University Press, 2001).

2. William Henry Perkin Jr. (1860–1929), with whom Chaim Weizmann worked, was the son of Sir William Henry Perkin Sr. (1838–1907), famous as the discoverer of the purple dye, mauve, and for other important achievements in organic chemistry.

3. The story about Chaim Weizmann and the Jewish homeland in Palestine is told in more detail in *Chemical Heritage* 20, no. 2 (summer 2002): 11–12, 26–29.

4. Bacon's *Essays* first appeared in 1597, with much-enlarged and rewritten editions in 1612 and 1625. There have been many edited versions, for example, Francis Bacon, *Essays*, ed. Michael J. Hawkins (London: Dent,1972).

5. Medawar's review originally appeared in *Mind* 70 (1961). It is reproduced in his *The Strange Case of the Spotted Mice, and Other Classic Essays in Science* (Oxford: Oxford University Press, 1961), pp. 1–11.

6. Alan Sokal and Jean Bricmont, *Intellectual Impostures* (London: Profile Books, 1998); U.S. title, *Fashionable Nonsense: Post-Modern Intellectuals' Abuse of Science* (New York: St. Martin's Press/Picador, 1998).

7. Other comments on Bohr's obscure lectures have been collected by Walter Gratzer in his *Eurekas and Euphorias: The Oxford Book of Scientific Anecdotes* (Oxford: Oxford University Press, 2002), pp. 51–52; the section on "The Lecturer's Craft," pp. 278–79, gives other examples of eminent people who lectured badly.

8. James Boswell, *Life of Dr. Samuel Johnson*, vol. 1 (n.p., 1775), p. 547.

9. Richard Dawkins, *The Blind Watchmaker* (London: Longmans, 1986; reprint, London: Penguin, 1988; reprinted with an appendix, 1991), pp. 37–40.

10. Richard Dawkins, *The Selfish Gene* (Oxford: Oxford University Press, 1976; reprint, Oxford Paperbacks, 1978, 1989), pp. 192–201.

11. A recent book by Robert Aunger, *The Electric Meme: A New Theory of How We Think* (New York: Free Press, 2002), does assume a physical reality for the meme, but the evidence for this does not seem persuasive to me.

12. Thirty-one generations ago Prince Charles (like all of us) had about 2.14 billion (2.14×10^9) ancestors. But in William the Conqueror's time the world population was only a few tens of millions, and the number of people in western Europe who might have been one of Prince Charles's ancestors was only a few million. The point is that many of his calculated 2.14 billion ancestors are the same people; in other words, he is descended from fewer people by multiple paths. It follows that those of us whose ancestors lived in western Europe are almost certainly also descended from William the Conqueror. The prince, however, can trace his ancestry, and most of us can't—perhaps that is just as well, since no doubt some of our connecting links were less than respectable, as were some of his.

13. *Nature* 22 (1880): 329. The short note on the subject is reproduced in Walter Gratzer, ed., *A Bedside Nature* (London: Macmillan, 1996). According to the note, the audience "enjoyed the singing about as well as if they had been placed in the upper circle of an ordinary Opera House." Unfortunately, no details are given of the techniques used or the size of the audience.

Chapter 7: Religious Belief

1. Richard Dawkins, *The Selfish Gene* (Oxford: Oxford University Press, 1976; reprint, Oxford Paperbacks, 1978, 1989), p. 198.

2. The book was called *Omphalos*, Greek for "navel," because Gosse was very concerned with the weighty problem of whether Adam, who allegedly was not born of woman, had a navel. Gosse concluded that, at creation, God included all the evidence for much greater ages for the Earth and the living creatures on it, so he would have given Adam a navel. In *Father and Son: A Study of Two Temperaments* (London: Heinemann, 1907), Philip's son, the distinguished poet and critic Sir Edmund Gosse (1849–1928), discussed this book and its reception, saying that "atheists and Christians alike looked at it and laughed, and threw it away."

3. The ideas of Lemaître were reported by one of the scientists present, distinguished physicist Victor Weisskopf, in his *The Joy of Insight: Passions of a Physicist* (New York: Basic Books, 1991). The Gosse and Lemaître stories are included in Walter Gratzer's *Eurekas and Euphorias: The Oxford Book of Scientific Anecdotes* (Oxford: Oxford University Press, 2002), pp. 154–56.

4. The sermon by Frederick Temple, preached in St. Mary's Church, Oxford, on Sunday, 1 July 1860, was published in *Christian Remembrancer* 38 (1860): 244.

5. The confrontation between Wilberforce, Huxley, and others was described in correspondence between several people present, and the accounts are by no means consistent. Excellent discussions of the debate in its historical setting are in Adrian Desmond, *Huxley: The Devil's Disciple* (London: Michael Joseph, 1994); Adrian Desmond and James Moore, *Darwin: The Life of a Tormented Evolutionist* (London: Michael Joseph, 1991); and Ronald W. Clark, *The Huxleys* (New York: McGraw-Hill, 1968).

6. A similar idea was proffered a millennium and a half earlier by St. Augustine, who suggested that the earth received from God the power to produce things *causaliter*, "of itself," without any specific divine act each time. This is quite consistent with the big bang theory and the theory of evolution.

7. Peter D. Ward and Donald Brownlee, *Rare Earth: Why Complex Life Is Uncommon in the Universe* (New York: Copernicus, 2000) argue persuasively that the conditions for the evolution and survival of higher life are so narrow and complex that they are unlikely to occur on more than one planet in the universe. It is hard not to reconcile this with the idea that the earth was designed for the existence of living things.

8. The word agnostic was coined by Thomas Henry Huxley in 1869. He commented later that "The Christian and the atheist were quite sure they had attained a certain "gnosis," . . . while I was quite sure that I had not." By *gnosis* he meant absolute certainty of knowledge. Atheists are sure that God does not exist; agnostics do not claim absolute certainty, but conclude that the weight of evidence is against it. Huxley's coining of the word is discussed in A. N. Wilson's *God's Funeral* (New York: W. W. Norton, 1999). Huxley coined the word in 1869 at one of the first meetings of the Metaphysical Society, which was active from 1869 to 1880.

9. David Hume's comment on the problem of evil, in his *Dialogues Concerning Natural Religion*, is as follows: "Is God willing to prevent evil, but not able?

Then he is impotent. Is he able but not willing? Then is he malevolent. Is he both able and willing? Whence then is evil?"

10. For more about the support of science by religious bodies, see Jack Morrell and Arnold Thackray, *Gentlemen of Science* (Oxford: Clarendon Press, 1981); and James R. Jacob and Margaret C. Jacob, "The Anglican Origins of Modern Science: The Metaphysical Foundations of the Whig Constitution," *Isis* 71 (1980): 251–67.

11. See Wilson, *God's Funeral*, p. 328.

12. Russell Stannard, *Science and Wonders: Conversations about Science and Belief* (London: Faber and Faber, 1996).

Chapter 8: What Is Truth?

1. For a general discussion of truth, see Felipe Fernández-Armesto, *Truth: A History and Guide for the Perplexed* (London: Bantam Press, 1997; reprint, London: Black Swan paperback, 1998).

2. See also Roger G. Newton, *The Truth of Science: Physical Theories and Reality* (Cambridge: Harvard University Press, 1997).

3. Lord Kelvin's opinion—by the end of the nineteenth century most scientific problems had been solved—was given in a lecture at the Royal Institution in London on 27 April 1900, and published as "Nineteenth century clouds over the dynamical theory of heat and light," *Philosophical Magazine*, ser. 6, vol. 2 (1901): 1–40. Similar opinions appeared in his book *The Baltimore Lectures on Molecular Dynamics and the Wave Theory of Light* (Cambridge: Cambridge University Press, 1904). The lectures referred to were given at Johns Hopkins University in 1884. This book was reprinted in R. Kargon and P. Achinstein, eds., *Kelvin's Baltimore Lectures and Modern Theoretical Physics: Historical and Philosophical Perspectives* (Cambridge: MIT Press, 1987). This volume also contains a number of essays on Kelvin's work in relation to modern physics.

4. The complete quotation from Winston Churchill's speech is "I cannot forecast to you the action of Russia. It is a riddle wrapped in a mystery inside an enigma."

Suggested Reading

Atkins, Peter. *Creation Revisited*. Oxford: Oxford University Press, 1992.

Aunger, Robert. *The Electric Meme: A New Theory of How We Think*. New York: Free Press, 2002.

Blackmore, Susan. *The Meme Machine*. Oxford: Oxford University Press, 1999.

Broad, William. *Betrayers of the Truth: Fraud and Deceit in the Halls of Science*. New York: Simon & Schuster, 1982.

Davies, Paul. *The Fifth Miracle: The Search for the Origin and Meaning of Life*. New York: Simon & Schuster, 1999.

——. *God and the New Physics*. New York: Simon & Schuster, 1983; Touchstone (paperback) edition, 1994.

——. *The Mind of God*. London: Viking, 1992.

——. *Are We Alone?* London: Penguin, 1995.

Dawkins, Richard. *The Selfish Gene*. Oxford: Oxford University Press, 1976; reprint, Oxford Paperbacks, 1978, 1989.

——. *The Blind Watchmaker*. London: Longmans, 1986; London: Penguin, 1988; reprinted with an appendix, 1991.

——. *River out of Eden*. London: Weidenfeld & Nicholson, 1995.

——. *Climbing Mount Improbable*. New York: W. W. Norton, 1996.

——. *Unweaving the Rainbow: Science, Delusion, and the Appetite for Wonder*. London: Allan Lane, 1998.

Ekeland, Ivar. *Mathematics and the Unexpected*. Chicago: University of Chicago Press, 1988.

Fernández-Armesto, Felipe. *Truth: A History and Guide for the Perplexed*. London: Bantam Press, 1997; reprint, London: Black Swan, 1998.

Gorst, Martin. *Measuring Eternity: The Search for the Beginning of Time*. New York:

Broadway Books, 2001. This book covers in an interesting way both the biblical and the scientific evidence for the ages of the universe, the earth, and life.

Gould, Stephen Jay. *Rocks of Ages: Science and Religion in the Fullness of Life.* New York: Ballantine, 1999.

Gratzer, Walter. *The Undergrowth of Science: Delusion, Self-Deception, and Human Frailty.* Oxford: Oxford University Press, 2001.

———. *Eurekas and Euphorias: The Oxford Book of Scientific Anecdotes.* Oxford: Oxford University Press, 2002.

Gribbin, John. *Almost Everyone's Guide to Science: The Universe, Life, and Everything.* New Haven: Yale University Press, 1998.

———. *In Search of the Big Bang: The Life and Death of the Universe.* London: Penguin, 1998.

———. *The Birth of Time: How Astronomers Measured the Age of the Universe.* New Haven: Yale University Press, 1999.

Hogan, Craig J. *The Little Book of the Big Bang: A Cosmic Primer.* New York: Copernicus, Springer Verlag, 1998.

Jones, Steve. *The Language of Genes.* London: HarperCollins, 1993; paperback ed., 1995.

Laidler, Keith J. *To Light Such a Candle; Chapters in the History of Science and Technology.* Oxford: Oxford University Press, 1998.

———. *Energy and the Unexpected.* Oxford: Oxford University Press, 2003.

Maddox, Sir John. *What Remains to Be Discovered: Mapping the Secrets of the Universe, the Origins of Life, and the Future of the Human Race.* New York: Simon & Schuster, 1999.

McKenzie, A. E. E. *The Major Achievements of Science.* Cambridge: Cambridge University Press, 1960.

Medawar, Sir Peter. *The Strange Case of the Spotted Mice, and Other Classic Essays on Science.* Oxford: Oxford University Press, 1961; paperback ed., 1996.

Menzies, Heather. *Whose Brave New World? The Information Highway and the New Economy.* Toronto: Between the Lines, 1996.

Newton, Roger G. *The Truth of Science: Physical Theories and Reality.* Cambridge: Harvard University Press, 1997.

North, John. *The Norton History of Astronomy and Cosmology.* New York: W. W. Norton, 1994. U.K. title, *The Fontana History of Astronomy and Cosmology.*

Rees, Sir Martin. *Before the Beginning: Our Universe and Others.* London: Touchstone, 1997.

Reeves, Hubert. *Atoms of Silence.* Cambridge: MIT Press, 1985; reprint, Toronto: Stoddart, 1993.

Reeves, Hubert, Joel de Rosnay, Yves Coppens, and Dominic Simonnet. *Origins: Cosmos, Earth, and Mankind.* New York: Arcade, 1998.

Schopf, J. William. *Cradle of Life: Discovery of Earth's Earliest Fossils.* Princeton, NJ: Princeton University Press, 1999.

Silk, Joseph. *The Big Bang.* New York: W. H. Freeman, 1989.

Sokal, Alan, and Jean Bricmont. *Intellectual Impostures.* London: Profile Books, 1998. U.S. title, *Fashionable Nonsense: Post-Modern Intellectuals' Abuse of Science.* New York: St. Martin's Press/Picador, 1998.

Stackhouse, John G. *Can God Be Trusted? Faith and the Challenge of Evil.* Oxford: Oxford University Press, 1998. This book presents a different point of view from mine.

Tanford, Charles, *Ben Franklin Stilled the Waves: An Informal History of Pouring Oil on Water with Reflections on the Ups and Downs of Scientific Life in General.* Durham, NC: Duke University Press, 1989.

Vanin, Gabriele. *A Photographic Tour of the Universe.* Willowdale, Ont: Firefly Books, 1998.

von Baeyer, Hans Christian. *Maxwell's Demon: Why Warmth Disperses and Time Passes.* New York: Random House, 1998. Paperback edition entitled *Warmth Disperses and Time Passes: The History of Heat.* New York: Modern Library, 1999.

Ward, Peter D., and Donald Brownlee. *Rare Earth: Why Complex Life Is Uncommon in the Universe.* New York: Copernicus, 2000.

Weinberg, Steven. *The First Three Minutes: A Modern View of the Origin of the Universe.* New York: Basic Books,1998.

Wills, Christopher, and Jeffrey Bada. *The Spark of Life: Darwin and the Primeval Soup.* Cambridge, MA: Perseus, 2000.

Wilson, A. N. *God's Funeral.* New York: W. W. Norton, 1999.

Wilson, Edward O. *Consilience: The Unity of Knowledge.* New York: Knopf, 1998.

Reading biographies of distinguished scientists is a useful and interesting way to gain an understanding of science. Biographies are usually written to be easily intelligible to readers not trained in science. The following are some biographies that relate to the subject matter of this book:

Bowen, Catherine Drinker. *Francis Bacon: The Temper of a Man.* Boston: Little, Brown, 1963.

Brown, Andrew, *The Neutron and the Bomb: A Biography of Sir James Chadwick.* Oxford: Oxford University Press, 1997.

Brown, G. I. *Count Rumford: The Extraordinary Life of a Scientific Genius.* Trowbridge, Wiltshire: Sutton, 1999.

Campbell, John. *Rutherford: Scientist Supreme.* Christchurch, New Zealand: AAS Publications, 1999.

Clark, Ronald W. *The Huxleys.* New York: McGraw-Hill, 1968.

Crowther, J. G. *British Scientists of the Nineteenth Century.* London: KeganPaul, Trench & Trubner, 1935. This includes an excellent article on Kelvin.

Desmond, Adrian. *Huxley: The Devil's Disciple.* London: Michael Joseph, 1994.

Desmond, Adrian, and James Moore. *Darwin: The Life of a Tormented Evolutionist.* London: Michael Joseph, 1991.

Goldman, M. *The Demon in the Aether: The Story of James Clerk Maxwell.* Edinburgh: Harris,1982.

Hall, A. Rupert. *Isaac Newton, Adventurer in Thought.* Cambridge: Cambridge University Press, 1992.

Henig, Robin Marantz. *The Monk in the Garden: The Lost and Found Genius of Gregor Mendel, the Father of Genetics.* Boston: Houghton Mifflin, 2000.

Keynes, Randal. *Annie's Box: Charles Darwin, His Daughter, and Human Evolution.* London: HarperCollins, 2001.

Macdonald, D. K. C. *Faraday, Maxwell, and Kelvin*. New York: Doubleday, 1954.

Maddox, Brenda. *Rosalind Franklin: The Dark Lady of DNA*. London: HarperCollins, 2002. An admirable biography that explains the work on the structure of DNA and the resulting controversy about priority in an easily understood way.

Smith, C. W., and M. N. Wise. *Energy and Empire: A Biographical Study of Lord Kelvin*. Cambridge: Cambridge University Press, 1989.

Thomas, Sir John Meurig. *Michael Faraday and the Royal Institution; The Genius of Man and Place*. Bristol, England: Adam Hilger 1991.

Tolstoy, Ivan. *James Clerk Maxwell: A Biography*. Chicago: University of Chicago Press, 1981.

Westfall, R. S. *Never at Rest: A Biography of Isaac Newton*. Cambridge: Cambridge University Press, 1980.

White, Michael, and John Gribbin. *Darwin: A Life in Science*. New York: Simon & Schuster, 1995.

Index

Page numbers in *italics* refer to illustrations.

221